MW01194347

# They Had NAMES

## Tracing the History of the North American Indigenous People

Nathaniel T. Jeanson

with illustrations by Cameron Suter

First printing: April 2025



Master Books, P.O. Box 726, Green Forest, AR 72638

Master Books® is a division of the New Leaf Publishing Group, LLC.

ISBN: 978-1-68344-424-4

ISBN: 978-1-61458-934-1 (digital)

Library of Congress Control Number: 2025934912

Cover by Diana Bogardus



Please consider requesting that a copy of this volume be purchased by your local library system.

Printed in the United States of America

Please visit our website for other great titles:

www.masterbooks.com

For information regarding promotional opportunities, please contact the publicity department at pr@nlpg.com.

*To all who have desired to know the pre-Contact history of North America*

## Praise for *They Had Names*:

[Nathaniel Jeanson] has indeed written a spectacular new and revised history of North American Indians, of the numerous Tribal Nations' origins and migration theories. There are definitely bridges between many great civilizations such as the Aztecs, the Longhouse Peoples and the Mound Builders, based upon their histories such as the Red Record and linguistic ties, as well as new scientific methods of tracing Y- Chromosome of DNA.

Helikinuva (Ted David Wilson),
Wakashan, Salishan, and Tsimshian First Nations

*They Had Names* is a remarkable historical reference not only for Indigenous peoples curious about the history of Turtle Island and when they arrived here but also for those who seek to understand the true historical migration of the peoples of this land…

Jeanson's narrative is interwoven with journeys to sacred Indigenous sites, which hold profound significance for the various nations of Indian people who lived centuries ago. Unlike the monumental structures of Greek civilization, the Indigenous peoples of North America left fewer physical markers, reflecting their deep respect for the land. They treated it with care, ensuring it remained as pristine as when they first visited. These sacred sites were places of ceremonies, gatherings, and other cultural practices for generations.

Jeanson's travels to these sites not only provide valuable insights but also moments of humor and self-discovery. His experiences offer glimpses into the meaning of these sites and the lives of the Indigenous peoples who inhabited these lands for centuries, imagining their historical journeys across the land.

Overall, *They Had Names* offers profound insights into the peoples of the land that many call Turtle Island, shedding light on their history through DNA analysis and other supporting documents. It challenges common assumptions and enriches understanding by presenting a detailed timeline of how, and perhaps when, the Great Plains of North America were settled. These revelations may surprise those who believed they already knew the region's history, whether through written, oral or graphic communication methods. He introduces another element in this analysis by using DNA data sets to help fill in some of the gaps.

The book's strength lies in its ability to bridge science with historical records. By utilizing DNA evidence and genealogical research, Jeanson provides a fresh perspective on the movements and settlement patterns of North America's Indigenous peoples. The findings are both surprising and enlightening, offering a clearer view of the past.

Though rich in scientific and historical data, the book remains accessible, blending technical analysis and research with engaging travel stories. Jeanson's personal reflections and anecdotes enhance the reader's experience, making this not just a study of history but a journey through it.

In conclusion, *They Had Names* is a commendable work that delivers a compelling analysis of a history that has long eluded many. Jeanson's dedication to uncovering the truth, paired with his approachable writing style, makes this book an invaluable resource for anyone interested in the origins and journeys of North America's Indigenous peoples.

Dana Soonias,
Red Pheasant Cree

Born and raised on the Navajo Reservation, I am proud to endorse Nathaniel T. Jeanson's journey to uncover the world in *They Had Names.*

I appreciate how Jeanson took the reader on an excursion through his research, allowing us to travel with him to different ancient ruins throughout the United States, walking us through his thought process, and, like a good guide, explaining the connection between his DNA research and tribal migration history.

For many Native Americans, our own history is a mystery to us. Yes, we have traditional stories of our migration to our heartland, but to translate those stories to current maps and timelines has not been composed in a way that is honest, respectful, and interesting—as Jeanson has been able to accomplish in *They Had Names.*

Sahvanna Benally,
Navajo

As a Cayuga nation member, Iroquois six nations, I have never come across a more clear and concise way of beginning to tell the story of Native American history. I'm thankful to God for the work Dr. Jeanson has done to start telling the true story of how Native America came about.

Seneca Van Aernam,
Cayuga Nation, Iroquois Six Nations

Prior to my full introduction to the Bible recently, I'm sure I would have discounted the timeline offered in the book. My new belief in the young earth theory makes the information here more relevant and answers a few other questions I had. By using DNA and linguistics Dr Jeanson was able to paint, for me, a better picture of the past.

Robert Rutherford,
Blackfeet/Potawatomi

This is a paradigm-breaking, mind-bending book!...

Jeanson is a scientist and scholar with a huge curiosity and the gift of quickly learning outside of his core discipline of molecular biology. He is also a synthesizer and a storyteller, looking to apply the principles of the scientific method to a very complex subject matter. This book is the result of his studies in DNA, genetics, historical and comparative linguistics, archeology, geology, cultural anthropology, western history, Native American culture, and Native American prehistory... .

Perhaps the biggest contribution Jeanson makes to the ongoing conversation about a more suitable narrative of the prehistory of North America, he re-opens the question of whether the *Wallam Olum* ("Red Record") is authentic or a fabrication. He argues persuasively that there are multiple places in the text of the *Wallum Olum* which are corroborated by historical records and archeological, biological, and linguistic data—data which could not have been known in 1836 when Constantine Samuel Rafinesque first published the entire text of the *Wallam Olum,* and his English translation of it.

This new evidence of the veracity of the *Wallam Olum* allows Jeanson to re-introduce significant pieces into the multidimensional puzzle of what happened in North America before the arrival of the Europeans. The *Wallam Olum* narrative helps to calibrate as never before the timeline of events in the history of the Lenni Lenape people leading up to their contact with the first Europeans. This time calibration can now be used to inform the developing revised narratives about the prehistory of other peoples.

Stephen M. Echerd,
Linguist and Language Revitalization Specialist

This book does what few before can claim: returns Native American histories to their owners, unifies the diverse Nations' stories across the continent, and offers a humble, yet thrilling, connection to the entire world.

Uniquely, this book stands as a beginning—not a conclusion—of a work. Reading it brought forward objections, excited agreement, and that-makes-so-much-sense moments. All of these appear welcome by the author, who invites you to contribute your own knowledge to this story.

If you are Native American, I encourage you to read this and compare to your own stories and experiences. You come away with a new appreciation for the amazing accomplishments and journeys of all the tribes, questions to ask the oldest elders, and a new connection to investigate of people once thought far away.

If you are a descendant of more recent immigrants to this county (in the last five centuries) this book is vital knowledge of the country you call home. It ignites a desire to know more and inspires you to see anew the landscapes of your own backyard.

Jeanson does the amazing by combining great storytelling with complicated science and his personal experiences. The beautiful mix will take you across America to places you never knew, to deep history and lost civilizations, and return you home with a longing to know more.

I recommend this book to any adventurous enough to consider that the forgotten may be found, that the different may be my brother, and that the ancient may speak from my backyard.

> Sarah Stewart grew up among Athabaskan and Yupik peoples and raised her kids among the Inupiat of Alaska

Jeanson's book on the history of the Native peoples of North America makes for a rare, unique, and interesting read. Even more so, since he presents it from a non-secular, biblical, young-earth creationist perspective, integrated with the gospel. Jeanson performed thorough research to present the true history of the Native peoples, encompassing historical, linguistic, ethnographic, and linguistic evidence. Jeanson's book intriguingly traces the movement of various Native tribes across the Americas. It is much like reading a detective novel. This book fills in a knowledge gap in the creation science literature.

Matthew Cserhati, PhD,
Associate Professor of Computer Science,
Concordia University Irvine

Who were the first Americans? Dr. Nathaniel Jeanson is on a mission to find out. His visits to Native American historical sites draws readers into a world of mystery where many times we've had more questions than answers. He then masterfully begins to answer those questions by showing how genetics confirms early accounts of Native American history. His level of discernment, attention to detail, and voluminous research in both indigenous history and genetics combine to provide a foundation for discovering the identity of the first Americans.

Georgia Purdom, PhD,
Vice President of Educational Content
and Director of Research, Answers in Genesis

This book represents a watershed moment in the creation-evolution debate. Creation scientists are now taking the lead in solving centuries-old scientific and historical mysteries. You'll be blown away and encouraged by how the true history in God's Word, combined with modern scientific techniques, is unshrouding history.

Ken Ham,
CEO, Answers in Genesis

# Table of Contents

# Acknowledgements

I am grateful, first and foremost, to God for the opportunity to explore the past in ways few people anywhere in the world or at any other time in history have been able to do. I recognize that the advances I've made are not because of my own cleverness or insight. They're due to the mercy of God giving me breath for each day and a sound mind to comprehend what I encounter. He has already given me far more than I deserve—the future hope of heaven, a clean conscience, and trustworthy promises of Scripture. Any discoveries I have made are the gracious and kind result of His enabling.

Many thanks to my wife and kids, who traipsed along on my Native American site excursions or held down the fort in my absence. My experiences on these trips will stick with me for a long time.

Thanks as well to my parents and my in-laws, both of whom were happy at a moment's notice to provide a place to stay in my travels around the country.

Cameron Suter has been an invaluable ally and collaborator. His efforts on the illustrations in these pages are just part of his contribution to this project. As a sounding board for visual communication and a wealth of ideas for making the visual point clearer, I'm especially grateful.

Walt Stumper, the now-retired librarian at Answers in Genesis, was a constant help in tracking down materials, books, and whatever I needed to get the answers I wanted on the pre-Columbian Americas.

Eric Dean was always prompt in getting me printed drafts of illustrations whenever I needed them.

Numerous lay and expert reviewers provided invaluable criticisms, corrections, and suggestions. I won't name them here, in part to protect their identities and careers. I thank you dearly for your work. Any errors that remain are my own.

A whole host of Native Americans and First Nation members have been wonderful contacts, sounding boards, and aids to the project at various stages. Some were early members of the study group. Others I've spoken with only briefly. So many were extremely helpful in educating me on Native ways and practices and were patient with my foibles, ignorance, and cultural faux pas. To protect their identities, I have also omitted their names. But do know how grateful I am to you for all your help.

I have had different supervisors at different stages of the project. Thanks to both Andrew Snelling and Georgia Purdom for your support and help. Thanks as well to the CEOs of Answers in Genesis, Ken Ham and Martyn Iles, for the opportunity to take this rewarding journey.

# 1

# Missing History

I've always known there was something *off* about the history of North America. I was born and raised in Wisconsin but grew up in two cultures. My dad's side supplied the American culture; my mother's family supplied the other. Because of her German roots, I spent nearly every summer of my boyhood years across the Atlantic visiting my German relatives and hitting as many tourist sites as we could. It was there, while traveling up and down the Old Country, that a gnawing feeling began to grow, a nagging sense that I was missing something profound about the history of my homelands.

I remember meandering in a boat down the Rhine River, medieval castles popping up on either side. We walked into one. Swords and armor lined the stone walls which rose high above us and obscured the sun. I felt like I had stepped back into the era of knights and maidens and jousts of the Middle Ages.

And then we'd return to the United States, and I'd remember that the Middle Ages came and went before any Pilgrims set foot in America. Curious.

One of our favorite German tourist destinations was on the Tauber River. *Rothenburg*, a stone-walled town, was quaint, colorful, romantic. The old-fashioned, half-timbered, and odd-shaped houses with low doorways crammed together astride narrow streets made it feel like it was locked in time, as if little had changed since its cathedral was built in the A.D. 1300s.

And then I'd come back to the United States and remember that Columbus sailed the ocean blue, not in 1392 but in 1492. Amazing. By the A.D. 1300s, European civilization had organized to such an extent that they were building architecturally stunning churches. But in the Americas, history was—about to begin?

As an adult, I returned to Europe, this time visiting Italy. Ancient ruins popped up everywhere in Rome, often right next to streets bustling with cars and lined with modern apartment buildings. I remember my host parking his car on a hill, me getting out, and then walking over to the edge of the vista to see vehicles below driving around one of the most famous first century A.D. sites, the Roman Coliseum.

It is jarring to walk among the remnants of a 2,000-year-old civilization, and then return to a land where the main historical attractions all date to the A.D. 1600s or later: Plymouth Rock, the Old State House in Boston, Monticello, Mount Vernon, Gettysburg.

And so, for years, I had wondered: In North America, where were the ruins of civilizations gone by, the evidence of millennia of history? Why was North American history so late, so short? What was I missing?

◂◂ ▪ ▸▸

Growing up, I had learned about the indigenous peoples here at European Contact—Algonquians like the Powhatan Confederacy in Virginia and like the Wampanoag in New England; Muskogeans like the Choctaw, Chickasaw, Creeks, and Seminoles who, along with the Cherokee, trudged the bitter path of removal from the southeast to west of the Mississippi; Siouans like the Lakota and Crow, who ruled the northern Great Plains; Cheyenne, Blackfeet, Arapaho, Kiowa, Comanche, Pawnee, Shoshone, and others who roamed the prairies and Rocky Mountains. Because I had grown up in Wisconsin, I had also been familiar with the indigenous peoples of the upper midwest: Menominee, Chippewa (Ojibwa), Winnebago (or *Ho-Chunk*), Fox, Sac, and Potawatomi. In school, I suppose we had also touched on the indigenous peoples of the southwest and northwest, nations like the Apache, Navajo, and Hopi, as well as the Nez Perce. But I had never learned the back story, the history of any of these peoples.

This omission was especially glaring when placed next to the detailed back story I had learned for the people who arrived on their shores—the Europeans. In high school I was taught that European civilization had been born millennia before Christ. It had sprung from the Minoans on Crete and from the Mycenaeans on the Balkan peninsula. I had learned that their linguistic relatives, the Greeks, had expanded their civilizational footprint in the first millennium B.C., sending out colonists far and wide across the Mediterranean world. I was told about Alexander the Great and his attempts to expand the footprint all the way to Central Asia. I had learned that the Romans, despite humble beginnings on the Italian peninsula, eventually overran the Mediterranean world around the time of Christ. I had learned about famous Roman rulers like Julius Caesar, Augustus, and Nero. I was taught that several centuries later, Rome was conquered by invaders. Barbarians from northern and eastern Europe, and Huns from Central Asia, ended Roman hegemony and ushered in the Middle Ages, eventually spawning the voyages of Columbus, of the Jamestown settlers, and of the Pilgrims.

And yet for North America, the Jamestown founders and the Pilgrims were the start of the narrative, not the end. I had been taught—again, in great detail—about the failure at Jamestown, about Plymouth Plantation, about the 13 Colonies, about the American Revolution, about the Louisiana Purchase and Manifest Destiny, and about the Wild West.

I wasn't taught the history of North America prior to European arrival.

For example, how long had the Wampanoag survived the cold winters of Massachusetts? Two thousand years ago, were they living on the Atlantic Seaboard? Were their ancestors even in North America? Or were they like the Huns in Europe, invaders from a faraway land?

What about the Seminoles? Had the Florida sun always shone on them? Or did the Seminoles once do battle with peoples living in the Caribbean? In Mexico? In Canada? In South America? Did they march their armies as many thousands of miles as Alexander the Great marched his?

What about the Lakota and the Crow? What nations, if any, preceded their rule of the northern Great Plains? Did they ever govern an empire as large as the one that the Romans commanded in Europe? Or was their political domain smaller, like one of the many Middle Ages-era kingdoms in Europe that popped up and disappeared in short succession?

My history classes hadn't answered any of these questions.

Thus, for decades, I had no ruins to point to, no millennia of history leading up to the A.D. 1600s; no sequences of kingdoms and battles; no cradle of civilization from which they had sprung; no names for the heroes and villains, the winners and losers of the ancient American past.

◀◀ ▪ ▶▶

And then...a glimmer of light began to creep through the opaque window to pre-Contact North America.

It started with little insights here and there. For example, sometime within the past nine years, I was strolling through the history section of our local library when Charles Mann's book *1491: New Revelations of the Americas Before Columbus* caught my eye. Here, finally, I thought, might be some answers to questions that had bothered me for years.

One of Mann's main conclusions contradicted the thinking from 60 years ago. Back then, the prevailing view was that, at the time of European Contact, Native American populations were sparse. Mann showed data suggesting that the Americas were brimming with people.

And then, 80% to 90% of the indigenous populations present in A.D. 1491 disappeared in the centuries following.

This sobering realization provided a stepping stone into another beam of light. About five years ago, I was trying to make sense of a puzzle that was pestering me at my day job. By training, I'm a biologist. At the time, I was working on the genetics of human origins. From DNA comparisons, I could reconstruct a putative family tree for global humanity—a tree that included branches connected to Native Americans.[1]

1 E.g., studies such as: Poznik et al. 2016; Karmin et al. 2015.

Back then, I wasn't seeking the pre-European history of the Americas. Instead, I wanted to understand the tree as a whole. But I was stumped on two questions: Where was the beginning—the root—of the DNA-based family tree? And how was the timescale of human history stamped on the tree? Answers to these questions were essential to my goals.

Mann's conclusions led the way to the solution. I knew that the Native American peoples had suffered a profound population collapse after European Contact. Where was the genetic smoking gun that 90% of the historical Americans had disappeared? This question pressed on me.

And so I pressed deeper into the data. It turns out that the population collapse is unmistakable—*if* you adopt a non-standard root for the tree and a non-standard timescale.

These discoveries didn't pull back the curtain on the sequence of historical battles and heroes in the pre-Columbian world. At least, not immediately. But they did set off a chain of events that would consume me for the next several years.[2] Along the way, I made two more unconventional discoveries—ones that finally brought me full circle to the pre-European history of the Americas.

The first discovery arose from a deeper analysis of the DNA of Native Americans. The prevailing scientific and historical dogma says that the Americas were settled once in the ancient past from Asia. The same dogma puts the migration in the incomprehensibly distant past—far before the dawn of human civilization.[3] The genetic research I was pursuing suggested that the Americas had been settled from Asia, not once but several times. It also planted at least two of the migrations squarely within the known history of civilization.

The second discovery was almost accidental. I don't even remember how it came about. But I do know the result.

---

2 I chased these data wherever they led and ended up following the implications for human history around the globe. In each region, the new root and timescale brought known history and DNA-based history into agreement, further confirming the validity of the unconventional approach that I was taking. In 2022, I published the results of this chase (see Jeanson 2022).

3 See Raff 2022 for discussion of mainstream debate over dates.

In mainstream science, the histories that Native peoples tell about their own origins and migration tend to be marginalized, if not dismissed outright. If this isn't immediately obvious to you, just think back to your own experience learning the history of the Americas. How much of what you learned was based on the standard scientific fields of archaeology, genetics, and linguistics? Then ask yourself how much the indigenous histories played a role in the conclusions that you were taught. If you can't think of a single example of the latter, or if you didn't even know that Native Americans have their own histories, then you've likely experienced the fruits of what I'm talking about.

As I dug into the DNA-based family trees for indigenous Americans, I found genetic confirmation of these Native accounts. One of the most explosive examples was the *Red Record*, or *Wallam Olum*. This document represented the putative history of the Delaware nation (the *Lenni Lenape*). According to the *Red Record*, the history was authored by the Delawares themselves.[4] In the early A.D. 1800s a Kentuckian named Constantine Rafinesque had brought the *Red Record* to the attention of the Western world. Rafinesque had obtained it from a physician, who had obtained it from a Delaware man.[5] For almost two centuries thereafter, the *Red Record* enjoyed occasional study and published analyses.

Then, in 1995, a graduate student in anthropology named David Oestreicher published his PhD thesis. Oestreicher claimed that the *Red Record* was a forgery, a hoax constructed by Rafinesque to garner himself publicity. Oestreicher's claims were so persuasive that the Delaware nation in Oklahoma "formally withdrew its earlier endorsement"[6] of the *Red Record*.

Around twenty-five years later, I was exploring the DNA-based family tree of humanity. It confirmed the history in the *Red Record*—

4 See the following for examples of English translations of the *Wallam Olum*: Brinton 1885; McCutchen 1993.

5 McCutchen 1993.

6 Oestreicher, D.M. "The Tale of a Hoax,", p. 23, in: Swann 2005.

directly contradicting Oestreicher's claims.[7] Questions, however, remained in my mind. So I dug deeper into the Native histories, including those from Latin America.

Time and time again, the Native histories and the DNA histories agreed, from the Incan's history of their arrival in South America,[8] to the Aztec's history of Central America,[9] to the histories from the linguistic relatives of the Delaware in North America.[10] This process has been so productive that, now, when I encounter an unfamiliar Native American history, my default position is to assume that it's true.

To clarify, I don't assume that every single recorded account of Native American history is accurate. My Native friends themselves have cautioned me in specific instances, telling me which 1800s ethnologists were diligent in transcribing Native accounts and which ones lacked discernment. But my attitude remains one of true-unless-proven-otherwise rather than the opposite.

◀◀ ▪ ▶▶

This book is the result of my multi-year deep dive into genetics and Native histories, as well as into North American archaeology and linguistics.

In this book, you'll discover pre-Columbian links between civilizations of the Old World—civilizations that we learned about in school—and civilizations of the New. You'll read about unexpected connections between Mexico and the tribes I learned about north of the Rio Grande. You'll also discover that ancient ruins from North American civilizations still exist, that these ruins tell dramatic stories, and that they can still be visited. Some of them might be in your own backyard. More importantly, I hope that, once you finish this book,

---

7 See Jeanson 2020. To be clear, I'm not claiming to have answers for all of Oestreicher's objections. Rather, given how closely the history in the *Red Record* matched the history implied by the reoriented DNA-based family tree, I find it hard to dismiss the *Red Record* as inauthentic. How could a forgery from the early 1800s anticipate genetic discoveries two centuries in the future? In my view, only real histories could do this, not hoaxes. See chapter 4 for additional lines of evidence. See also Appendix A.

8 See Sarmiento de Gamboa 1572. The straightforward reading of the timescale in this document implies an arrival in South America precisely in line with the genetics I show in chapter 3.

9 See Bierhorst 1992. The straightforward reading of the timescale in this document implies an arrival in Central America precisely in line with the genetics I show in chapter 3.

10 See chapters 4 and 5 and Appendix B.

you'll walk away with a profound sense of how dynamic the Americas were before European Contact. Here, on this continent, peoples and kingdoms rose and fell, great battles were fought, heroes were enshrined—heroes whose names have been preserved.

You won't find the complete history of pre-Columbian North America within these pages. We have the most historical information for the millennium prior to European Contact, especially the latter half of that millennium, but for the thousands of years prior, our information is still fragmentary. I've divided this book into two parts to reflect this temporal asymmetry in our knowledge.

I'm guessing that you'll find the pace for Part I to be slower than the pace for Part II. The events during Part I lack the ethnic detail and personal angle of Part II. There's more archaeology and less connection to modern tribal names. There are also more unresolved questions. Part II tends to focus more on the story, relying on the scientific conclusions reached in Part I. Part II is also full of histories connected to specific tribal peoples. Part I sets the framework for Part II, so if you persevere through Part I, I think you'll find Part II to be richly rewarding.

You might be wondering why I would even write a book, given the incompleteness of our understanding of the ancient past. Why not wait until more research is done and more results are gathered? For one, I think the information we currently possess is revolutionary and worth sharing. For another, I hope that, if nothing else, this book will inspire others to pick up the baton and run with it, to enlarge our understanding of and appreciation for all those who came before. Those who catch the vision for this work might eventually make discoveries that contradict some of what I've written in this volume. In fact, if this doesn't happen, I would be surprised. Science, by nature, is uncertain and always in flux. This book represents research happening right now. I fully expect some of the conclusions I've made to change with time.

I've written this book primarily with a lay audience in mind, not my nerdy professional peers. If you find yourself in the latter group, or if you're just looking for more technical justifications for my conclusions, or if you have burning technical questions (e.g., "Why doesn't he use ancient DNA in his analyses?") and would like the detailed science behind what I claim, see **Appendix A** in the back of this book.

I've also tried to bump many of the specific technical justifications or explanations for my points to **footnotes**.

For everyone reading this book, I recommend that you go to the online map of North America at Contact.[11] In subsequent chapters, especially in Part II, we'll explore the history of most of the bewildering array of peoples and nations that called North America home. Many of these nations might be unfamiliar to you. If you acquaint yourself with the map now, and if you keep it handy as you're reading, it will help keep the amount of detail in the coming chapters from feeling overwhelming. For both lay and technical audiences alike, I hope that, by the time you finish this book, you'll look at North America—and the peoples still in it—in a radically new way.

◂◂ ▪ ▸▸

## A word to my Native American friends

If you're reading this book as a Native American, and if you're wondering why a European-American would try to tell the story of Native American history, I'm glad you asked. To be clear, I don't pretend to have any Native American ancestry. Nothing in my family tree suggests that I have any genealogical links to the indigenous peoples of the Americas. In fact, I've spent nearly all my life swimming in a thoroughly Caucasian world.

So why write on Native American history?

Because I don't want my fellow Caucasians to live like I did—ignorant of those who came before and unaware of those who remain. How can someone esteem the indigenous heroes of North America if the pre-European history of North America is never taught? How can someone respect what they don't know?

To be clear, I'm not proposing that Caucasian-Americans learn pre-Columbian history and then embark on a grand plan to act like "white saviors" to the Natives who are still here. I'm not even proposing that Caucasian-Americans do anything but leave you alone. Rather, my ambition is to restore the rightful respect and appreciation for the peoples who were here before Europeans, as

11 <answersingenesis.org/go/theyhadnames/>

well as for their descendants. I want my kids to grow up knowing, appreciating, and respecting the Native peoples.

This may not be an earth-shattering goal, but I think it's a worthy one. I grew up without knowing and without appreciating. I don't want others to repeat my path.

I know that the U.S. government has a long and sordid history with Native Americans. I'm relieved to say that I'm not with the government. My current employer, Answers in Genesis, is a Christian, creationist non-profit. They are also not with the government. I don't receive any government money for what I do. In fact, because my research findings are unconventional and challenge the mainstream narrative, I will likely never receive any government money.

I recognize that this book uses scientific methods that have caused angst, hurt, and even anger in the past. As you read, you'll find plenty of discussion of archaeology and genetics, as well as linguistics. I know that some Native Americans are concerned that these tools will be used, for example, by the government, to take away their lands, their loved ones, and their identity. However, because this book is one of the rare examples of science confirming indigenous histories, my hope is that what you read within these pages will be a help, and not a hurt, to the Native American community. If nothing else, I hope that this book returns to your community the dignity and respect that you and your ancestors are due. Fellow human beings deserve nothing less.

## A word to my young-earth creationist friends

If you're reading this book as a young-earth creationist, this section is for you. You might be wondering: "Why should this book matter to me?" I'm glad you asked.

My answer is succinct: This book pushes back against one of the longest-standing objections to young-earth creation science. "What objections?" For decades, critics of creation science have made specific demands. They flow from the nature of science:

> *The most important feature of scientific hypotheses is that they are testable* [emphasis theirs].[12]

In other words, science—as a discipline—lives in the perpetual present. All claims are provisional. They are subject to ongoing experimental evaluation. The best scientific claims are those which survive the most attempts to disprove them. For years, evolutionists have insisted that creationists should publish—and then experimentally test—scientific predictions.

Happily, for young-earth creationists, the conclusions in this book are based on more than a decade of a consistent pattern of making testable genetic predictions, and then seeing these predictions fulfilled by later experiments. These have led to more predictions and more fulfillments. I've also witnessed these genetic findings match data from other fields, whose own data harmonize across disciplines. In short, my conclusions are working well.

To put it bluntly: This book represents a monumental reversal in the history of the creation-evolution debate. As a participant in this debate, my book is not primarily a refutation of evolutionary claims about pre-Columbian North American history. Instead, this book represents a replacement of those claims with a superior paradigm.

For more details and documentation on this paradigm shift, see **Appendix A.**

12 Futuyma and Kirkpatrick 2017, p.578.

# Part I

# Unexpected Journeys

# 2

# The First Americans
## 1000 B.C. to A.D. 300s

Say *Native American* or *American Indian*, and my mind immediately conjures several detailed images and scenes. I'm sure yours does, too. Perhaps you picture a vast treeless prairie with a dense sea of brown bison thundering in unity in one direction as bare-chested Native Americans on horseback rapidly string arrows to their bows and pick off individual animals in quick succession.

Maybe you picture the same treeless prairie, empty but for a group of tipis where another set of Native warriors with painted chests and feathered headdresses perform a war dance around a roaring fire.

Or perhaps you can see in your mind's eye the dense deciduous forests of the eastern United States in which hunting parties armed with bows and arrows and clad with buckskin leggings and moccasins slide noiselessly through the tall grass and dense underbrush in pursuit of deer. You might also picture their homes—clusters of wigwams near fields of corn, beans, and squash.

We have these mental pictures because these are the images that European settlers gave us from their initial encounters with Native Americans, or from interactions thereafter. The images are telling, more so for what they don't say than for what they do. Missing from these images are depictions of scenes known only from archaeology.

It turns out that the beginning of Native American history looked quite different from the scenes at Contact. The ruins of these ancient cultures still exist. These ruins also happen to be the only link that we have to this early history. Because in the A.D. 300s, the first Native American world disappeared.

◂◂ ▪ ▸▸

I had my first aha moment about the significance of these ruins almost by accident. On a whim one September weekend in late summer of 2023 as the leaves were beginning to change, my wife, my four small children, and I went on a family outing to *Fort Ancient*—a pre-European Indian site off I-71 that connects Cincinnati and Columbus.

After exiting I-71 and following winding state highways for about four miles, we saw the sign for Fort Ancient in a grassy field, pulled off, and parked to the left of the main building on the premises. As we got out and walked toward the small museum, we immediately spotted a non-descript mound before us. The hill was only a few feet in height, and it was covered with stones. It didn't seem like much so we moved on into the museum, paid the fee, and walked through.

It took me about an hour before the significance and size of Fort Ancient finally began to sink in. I learned that most of the Fort Ancient site dates back to around the time of Christ. The *Hopewell* culture constructed it. The main ruins are a series of rising and falling earthwork walls, some of which are 23 feet high.[1] The site itself is 126 acres.[2] It sits on a plateau, and at the far southwestern end of the complex, the view over the bluff is lovely.

But it wasn't the size of the site that most arrested my attention. It was discovering that the site was built with astronomical precision. Some of the mounds and gaps in the walls are aligned to the summer and winter solstices.[3]

I immediately thought of the ancient Maya, who lived far to the south in Mesoamerica.[4] They were also expert astronomers and kept a

1 https://www.ohiohistory.org/visit/browse-historical-sites/fort-ancient-earthworks/
2 https://historicsites.ohiohistory.org/fortancient/
3 https://www.ohiohistory.org/visit/browse-historical-sites/fort-ancient-earthworks/
4 A geographic term referring to most of Mexico and to parts of Central America.

precise calendar. It just so happens that the beginnings of the Classic Maya era were contemporary with the Hopewell. Interesting.

◂◂ ▪ ▸▸

I knew from books that the Ohio area was an epicenter of Hopewell activity. After my visit to Fort Ancient, I learned that some of the best sites were all within a 2- to 3-hour drive of my house. So I waited for a sunny, balmy forecast. Then, in early October, I set out to visit them.

East of Columbus, I glided off state highway 16 onto state highway 79. I was in Newark, Ohio, a town of just over 50,000 people. A mile later, the Newark Earthworks parking lot popped up on my right—smack in the middle of town. I parked, got out, and started walking north up the sidewalk through what looked like a park. A gently sloping grassy hill rose on my left.

At the top of the small incline, the grassy hill on my left dipped down into a ditch. The ditch rose again and met about 50 feet of level ground before meeting another ditch, which rose to another grassy hill. After a stop in the visitor center, I walked between hills and entered an expansive grassy field.

The two hills were actually part of a single ring—the Great Circle—that rose above eye level (**Figure 1**—see the eyeball-like shape in the lower center of the map). When I had initially parked, I wasn't able to see into the ring. Once I was in the enclosure, I wasn't able to look out over the walls to know who was immediately on the other side (**Figure 2**). I felt almost like I was in a stadium or an amphitheater.

Walking the perimeter, I realized the size of the Great Circle. It's over half a mile to skirt the ditches along the inside ring of the wall. The footprint of the circle encompasses 30 acres.[5] I felt it as I hoofed my way around, trying to take in the whole structure, knowing I had several more sites to visit that day.

In the middle of the circular enclosure, a small set of hills rose together. Their purpose was unstated. But I had a guess.

When I walked the grounds that day, inside the Great Circle was quiet. But it wasn't hard to imagine a very different picture when this

5 https://www.ohiohistory.org/visit/browse-historical-sites/newark-earthworks/

monument was in use before Europeans arrived. I could easily picture a sea of people inside the Great Circle, surrounding the central hills where leaders conducted a massive ceremony or gave a motivational speech.

I exited the enclosure the same way I entered (**Figure 3**). In front of me were low earthen walls on either side, as if to delineate an ancient path (**Figure 1**—see lines extending to the right and up from the eyeball-like shape). Then the causeway abruptly stopped. Like so many ancient earthworks, time and modern construction have obliterated much of what once stood. (**Figure 1** represents the state of the site in the A.D. 1800s.)

About half a mile away as the crow flies, in a residential district at the end of a small side street and adjacent to what looked like an industrial park, the small corner walls of what was once a square enclosure (**Figure 1**—see the square near the center of the map) rise about one or two feet from the ground, right next to someone's backyard. This enclosure was once connected to the Great Circle via an earthwork-lined walkway (**Figure 1**), but no more.

A two-mile drive from the corner wall plot wound me through residential and commercial streets to the Moundbuilders Country Club, the last earthwork stop for me in Newark. Here, like at the Great Circle site, earthwork walls rise and obscure the view inside the enclosure.

Unlike the first circle that was open to the public, the Country Club site was member-access only. A small platform next to the parking lot lets the average visitor peer over the walls inside the earthwork. I could see about half of a large circle, a few walls of an octagon, and a short-walled causeway that connected the two. All three structures have been fully preserved (**Figures 1, 4**).

Despite not being allowed to walk through the earthworks, I was still struck by their massive scale. Had I walked the inside perimeter of both the circle and octagon, I would have traversed almost a mile and a half. Combined, the area enclosed is 70 acres.[6]

Again, the atmosphere was quiet. I saw no one else climb the platform to survey the mounds while I was there.

Two millennia ago, it may have been just as quiet:

6 https://www.ohiohistory.org/visit/browse-historical-sites/newark-earthworks/

> The Octagon is aligned with the lunar cycle and every 18.6 years the moon rises directly through the passageway linking the Observatory Circle and the Octagon.[7]

Perhaps hushed silence once fell over onlookers who stood in the observatory and watched the lunar rise mark the passage of another cycle (**Figure 5**).

Who built these astronomically and geometrically precise structures? Who were the Hopewell people? Relatives of Mesoamerican peoples? I didn't have answers at the time, but my day's journeys weren't done.

I left Newark and drove an hour and twenty minutes southwest to two sites along the Scioto River in the town of Chillicothe. Right off a major highway sits the more famous of the two, Mound City Group. It abuts the western side of the river. Low walls form a rounded circle making a footprint of almost 17 acres.[8] Inside the enclosure, and visible to anyone outside of it, are an irregular arrangement of 25 mounds, used for burial (**Figure 6**).[9]

After walking the grounds of the Mound City Group, I drove 10 minutes across the river to Hopeton Earthworks. The walking path led up a rise. From the vista, I saw what initially appeared to be nothing but farm fields. However, I could see the long grass outlines carefully cut to mark the spots where several earthworks once stood—a couple of circles and a square-ish shape (**Figure 7**). In the distance, a long corridor diagonal to the intersection of the circle and square trailed off.

Once again, builders of this site had the heavens in view:

> The parallel walls were constructed to align with the sunset on the winter solstice.[10]

The trails didn't permit me to walk the grounds where the earthworks once stood, but the scale was, nonetheless, apparent. Together, the circle and square covered 37 acres.[11]

---

7 https://www.ohiohistory.org/visit/browse-historical-sites/newark-earthworks/
8 https://www.nps.gov/hocu/learn/historyculture/mound-city-group.htm
9 https://www.nps.gov/hocu/learn/historyculture/mound-city-group.htm
10 https://www.nps.gov/hocu/learn/historyculture/hopeton-earthworks.htm
11 https://www.nps.gov/hocu/learn/historyculture/hopeton-earthworks.htm

A 15-minute drive to the west of Hopeton was the biggest site yet. I pulled into the parking lot where the tree line obstructed the view of the Hopewell Mound Group site. Walking north along the sidewalk path, I stopped where the tree line parted, and a sign directed my attention to a vast field of mowed grass to the west. It was hard to know what I was looking at. The length of the field was more than 13 football fields. Nothing caught my eye except gigantic power lines in the distance cutting through the middle of it.

I continued north along the trail where the tree line resumed and again blocked my view of the field. The path eventually turned west, but I still couldn't see the site for all the foliage. Above my head was open sky, but to my left the trees still impeded my sight.

The trail began to rise, and, about halfway along, the tree line broke again, this time to let the powerlines through. A sign marked a place where I could gaze onto the massive field about 40 or 50 feet down below. Again, nothing in the field caught my eye. The break in the tree line struck me more as a scenic overlook—onto a pretty (albeit empty) grass field.

The trail continued to meander west, eventually entering the woods. At this point, earthwork walls finally began to appear to my left, rising and falling with trademark regularity. Then the path led down and curved south.

In a few minutes, I emerged at the grassy field, at the opposite end of where I first viewed it through the break in the trees by the parking lot. My starting point now looked very far away. But the short grass connecting the two points showed the beginnings of a path. I followed it.

From where I entered the path, the distance was too great and the cut grass outline too fuzzy to know where the path would end. I could see signage at irregular places in the field. Following the path, I ended up at each one.

Apparently, the field had once contained multiple mounds. Very low rises were visible—if I looked hard. But nothing in the visible remains was extraordinary. In fact, if I had driven down the road that paralleled the field, I think I would have hardly given the field a second glance. It looked like a farmer's field with natural rises and falls; nothing seemed out of the ordinary.

It wasn't until the path reached an area close to the powerlines that the dots finally connected for me. I looked back up toward the vista where I had stood at what I thought was just a scenic overlook. At that point, I realized the vista wasn't just a sightseeing spot. Earthwork walls once sat on top of it. From the map, I knew the full circumference of the enclosure. The walls had once formed a rough D-shape connected to a small square (**Figure 8**). I was standing in 130 acres of what once must have felt like a gigantic arena.

And yet…the entire Hopewell complex could have been bigger than anything I imagined that day:

> Bradley Lepper…has offered one of the truly new and testable hypotheses regarding the Ohio Hopewell archaeological record. He has found a number of historic records and some aerial images that suggest the parallel walls that begin at the Newark Earthworks stretch for at least 60 miles (96.5 km) to the south-west, and likely connected Newark with the great ceremonial centers of the Scioto River valley near Chillicothe.[12]

The Hopewell economic reach was even greater:

> One of the most impressive characteristics of Ohio Hopewell archaeology is the variety and vast quantities of artistic objects that are made from raw materials that are not found in the Ohio River valley. Particularly notable are obsidian and grizzly bear teeth from the Rocky Mountains, copper from the northern Great Lakes, mica and quartz crystals from the Appalachian mountains, and a range of shark teeth, barracuda jaws and conch shells from the coast of Florida. … As scientific data became available indicating how far these items had travelled, the general assumption developed that the items were obtained through trade. … But the paucity of exotic items like obsidian at archaeological sites between the Rocky Mountains and Ohio have led many scholars to abandon the idea of trade in favor of more long-distance movement of people.[13]

---

12 Lynott 2014, p. 86.
13 Lynott 2014, p. 25.

On a map (**Figure 9**), the geographic scale of this 2,000-year-old network almost boggles the mind.

You'll notice that the map (**Figure 9**) does not connect the Hopewell to Mesoamerica. Even though the Hopewell and Maya were contemporaries in time, no archaeological connections between them have been discovered—yet.

Also, unlike the Maya, the Hopewell didn't live in massive cities. Instead,

> Hopewell people lived, for the most part, in small, isolated communities, sometimes comprising no more than one or two extended families.[14]

Still…these "isolated communities" of Native Americans left profound testimonies to their astronomical, mathematical, and engineering prowess.

◀◀ ▪ ▶▶

Far to the south of the Ohio Hopewell, in what is now Louisiana, the *Poverty Point* culture arose (**Figure 10**). Like the Hopewell, they also built earthworks—both mounds and geometric structures (**Figure 11**). In fact, the people at Poverty Point were so industrious that they built the second largest earthwork in the eastern United States.[15]

Like the Hopewell, the constructions at Poverty Point were astronomically aligned—to the equinoxes rather than to the solstices or cycles of the moon.[16] They were also mutually well-connected:

> For a short period the Poverty Point site and nearby lesser centers were the nexus of a vast exchange network that handled great quantities of exotic rocks and minerals, such as galena, from more than ten sources in the Midwest and Southeast, some of them from as far as 620 miles (1,000 km) away.[17]

Unlike the Hopewell, the people at Poverty Point lived, not around the time of Christ, but a millennium earlier, circa 1000 B.C., and lasted

14 Fagan 2019, p. 259.
15 Kidder, T.M., "Poverty Point," p.462, in: Pauketat 2012.
16 Fagan 2019, p. 197.
17 Fagan 2019, p. 197.

only three centuries.[18] With respect to Mesoamerica, this puts Poverty Point much earlier than the Classic era of the Maya. In fact, it makes them contemporaries of the founders of a cradle of Mesoamerican civilization. The Olmecs, the sculptors of the unforgettable giant stone heads (**Figure 12**), founded their society on the Gulf of Mexico (**Figure 13**). We also have hints that the Olmecs were the inventors of the complex calendrical system used by the Maya.[19] Olmecs also happened to build earthen mounds (**Figure 14**). Given the nearness of the Gulf to Louisiana (**Figure 13**), it's hard not to wonder if the Olmec and the people at Poverty Point were related.

◂◂ ▪ ▸▸

Again, the pictures I just painted aren't descriptions of contemporary, thriving Native American cultures at the time of Contact. Early European settlers didn't convey the process of ongoing geometric earthwork construction. They didn't talk about astronomical observations on the part of the living Native Americans to measure and map the 18.6-year cycles of the moon, or the equinoxes, or the solstices. The settlers, upon arrival, didn't describe anything like the Hopewell or even the Poverty Point cultures taking place.

The reason? In the A.D. 300s to 600s, something changed in North America. Something profound.

---

18 Fagan 2019, p. 196-197.
19 Coe and Koontz 2013, p. 79.

# 3

# The Event
## A.D. 300s to 600s

On April 8, 2024, I found myself cruising through the rolling farmlands of southern Wisconsin towards an Indian mound destination across the Mississippi River in Iowa. I knew from textbooks that the southern Wisconsin area and its borderlands (**Figure 15**) were once the homelands of a mound culture (**Figure 16**) that was different from Hopewell. I wanted to see the ruins of the *Effigy Mound* culture for myself.

More importantly, I wanted to understand the relationship between Effigy and Hopewell. Current thinking puts Effigy between A.D. 700 and A.D. 1100.[1] Just before this period, in the A.D. 300s, Hopewell disappeared.[2] I wondered if there might be a cause-effect relationship.

Just before I reached the Mississippi River and crossed into Iowa, I noticed the topography change. The hills became steeper and much more pronounced. Instead of field after field on all four sides of me, I was suddenly looking up at bluffs, with the road taking me through a valley between them. On the Iowa side of the Mississippi, the bluffs were steep and dangerous, with sheer cliffs and falling rocks.

The Effigy Mounds National Monument is broken into two sections—a north unit and a south unit. The trail from the visitor center

1 Birmingham and Rosebrough 2017, p. 154.
2 Fagan 2019, p. 266.

zigzagged as it took me steeply and quickly to the top of the bluffs on the north unit.

Even though I made a beeline for the most distant mounds, the trail still took me past several sets of other mounds. One of the first mounds I saw didn't look like an effigy mound at all. It was a series of low, conical-shaped mounds connected by low log-shaped rises into *compound* mounds. Their height was rarely more than a foot or two. In fact, were it not for the provided map, I likely would have missed several on the hike. I didn't know what to make of these ruins.

Then I saw an actual effigy mound—the *Little Bear Mound*. It looked like the conical mounds, except that I could kind of make out the shape of a head, body, and legs. Again, the height of the mound was so low that it was hard to see. Also, the anatomical detail was minimal. "Four-legged" would be an equally accurate description, without needing to specify the actual animal. "Bear" fits, mainly because of the head shape and lack of much of a tail. But the mound certainly didn't attempt to draw the viewer in like a fresco or painting might. I moved on.

Just a short distance farther up the trail, I encountered the *Great Bear Mound*. At 70 feet by 137 feet,[3] it impressed me more than any of the other mounds I had seen thus far. In fact, I found out that it's the biggest single effigy mound in the entire park. Like the Little Bear, the Great Bear has a recognizable head, body, and legs. Also like the Little Bear, there's not much more to its shape.

None of these mounds produced the same sense of awe in me that the Hopewell mounds did. There were no high walls, stadiums, or amphitheaters. I had also read that scholars failed to find any compelling astronomical correlations with these earthworks—no alignments to solstices or equinoxes or lunar cycles.[4] If nothing else, it was clear that effigy mounds were different from the Hopewell mounds. Perhaps built by a different people.

As I continued along the trail, I kept asking myself, "What was the purpose of the effigy mounds?" Burial, for one. Archaeological excavations have revealed this facet. But why form their graves in the shape of animals? Was there no other angle to understanding this

3 As per the brochure.
4 E.g., see Birmingham and Rosebrough 2017, p. 149-151.

culture? What was the point? Who were these people? I knew from reading textbooks before my trip that the academic community was undecided. My initial observations produced little insight.

I continued following the trail until I reached the next-to-last set of mounds. They were two *linear* mounds, shaped like low pills. I stood next to them in an elevated area as I looked at the map to see where the trail would take me next. The last leg wound almost in a circle before taking me to Hanging Rock. I was a bit befuddled. From where I stood, it seemed like a straight—and short—shot to the mounds. Why take me away and around, away from Hanging Rock? Just to give the visitor more exercise? Having already hiked a couple of miles, I was miffed.

And then I found the answer. The Hanging Rock mounds sit on a cliff overlooking the Mississippi. On three sides, the only way up is via rock climbing or via movement akin to climbing a ladder or stairs. Only the north side allows the visitor to hike a manageable dirt path up to the top.

I followed the narrow trail up to the mounds themselves. Directly in front of the mounds sits a bench, at a slightly lower elevation than the mounds themselves. Because the site sits on a bluff, the view beyond the mounds was almost exclusively blue sky. I'm afraid of heights, and so I found the view terrifying—terrifying but manageable. In other words, it was spectacular. I didn't want to leave.

*What a way to honor your ancestors*, I thought. Imagine carrying dirt up all this way just to bury someone. Or, even if there was enough dirt up here, imagine working at this height and in this danger just to show respect to your deceased family. To me, the message was clear. This impressed me more than anything else I had seen on the trail.

And then I came to another realization. Up to this point, I hadn't bothered to read the brochure. Now I did, and I learned that Effigy Mounds National Monument was actually a park containing mounds from several cultures besides Effigy. In particular, it was thought that the conical mounds were centuries, if not a millennium, older than Effigy. On Hanging Rock, the mounds were conical. In other words, the most impressive location yet was picked, not by the Effigy Moundbuilders but by a culture that preceded them.

As I followed the north unit trail back, this time visiting all the side trails, a pattern began to emerge. All the best overlooks and bluffs seemed to have conical mounds in the positions of prominence. Effigy mounds had been built farther back from the cliffs—consistent with Effigy having arrived later.

Late in the afternoon, I drove the short distance from the north unit to the day-use area to the south. Unlike the north unit, the south unit ascent wasn't a zig-zag. It was straight up. My leg muscles didn't take long to protest.

At the top, though, the scenery was well worth the climb. The trail to the main and most distant attraction, the *Marching Bear Group*, was beautiful. On either side, the ground sloped down quickly. It wasn't a cliff, but it made me feel like I was walking on the top of a ridge. Lots of blue sky to the west (my right). As the path headed due south, the sinking sun cast lovely shadows through the trees.

At the Marching Bear Group, I received a rich reward. About ten bear-shaped effigy mounds and three bird-shaped effigy mounds are arranged in a pattern that the park staff said looked like the Little Dipper constellation. I could see what they meant.

Though the skies were echoed in the mounds, the arrangement on the ground itself followed what scholars have already noted: Effigy mounds tend to follow the natural topography.[5] In this case, the rule produced a lovely effect. The Marching Bear Group sits on the downward slope of a Mississippi-facing ridge. The mound area itself is clear of trees, but trees still line the area to the south, north, and west. The east drops off, not like a cliff but steep enough to produce a beautiful vista toward the Mississippi. The Effigy Moundbuilders, like the conical moundbuilders at the opposite side of the park at Hanging Rock, picked a great spot to honor the deceased.

The Marching Bear Group didn't share the slope with any conical mounds. However, on another shelf below and east of them toward the Mississippi, conical or compound mounds sat on a bluff. I had to follow a separate side trail to reach them. I guess it's possible that the latter were built first, and then the Effigy Moundbuilders came in later and picked a spot farther back but higher up.

5 Birmingham and Rosebrough 2017, p. 125.

Elsewhere on the south unit, the *Compound Mound Group* sits on a peninsula oriented along a north-south axis. The path approaches the Group from the south. The north end terminates in a bluff. Going backwards from the bluff—the position of prominence—the first mounds to appear are compound mounds. Farther south along the trail, two effigy mounds—a bird and bear—appear, as if they came later.

During my time at Effigy Mounds National Monument, I never saw any Hopewell mounds. I was never able to compare Effigy and Hopewell directly. But where I did see Effigy, I always found another older mound nearby.

Back home in Kentucky, I learned from the academic literature that this pattern is generally true. Effigy mounds tend to occur in groups with other types of mounds.[6] How interesting. If nothing else, it seemed like the builders of Effigy were trying to make a statement. Clearly, they were different from earlier cultures. Yet the Effigy people seemed to feel a need to connect their culture to those who came before—either to gain legitimacy or to assert dominance.

◂◂ ▪ ▸▸

Did Effigy rise because Hopewell fell? Did builders of Effigy cause the fall of Hopewell—as if there were two groups at odds with each other? If Effigy overthrew Hopewell, did Effigy then try to assert their dominance by building mounds at old sites that previous people had revered? All of these questions were ultimately wrapped up in a bigger one: *Who built Effigy?*

For *who* questions, I've found that the best clues are drawn, not from the visible realm of archaeology, but from the invisible realm of DNA. DNA is the *stuff* of heredity, the biological instruction manual for making each person's physical features, from eye color to hair color to eye shape to height and skin tone. Because DNA is passed from one generation to the next, it forms the basis for reconstructing family trees.

To make a long story short, certain types of DNA lend themselves better than others to building family trees. The DNA that we inherit from both parents is good for finding close relatives. For digging deep into the genealogical past, we need to analyze the DNA that we inherit

---

6 Birmingham and Rosebrough 2017, p. 120.

from only one parent. Mothers pass on *mitochondrial DNA*; fathers pass on *Y chromosomes*. The latter produces the best, most precise family trees.[7]

What do the Y chromosomes of Native American males reveal about the history of North America? For the recent past, the answer is *almost everything*. In fact, before we can infer the pre-Columbian history of Native Americans, we have to first solve a practical problem. In the last few (post-Columbian) centuries, European colonialism and the awful practice of the Trans-Atlantic slave trade have brought European and African DNA to the New World. Today, some Native American men belong to branches on the family tree that lead back to Europe and Africa. We must wade through and remove these branches from the tally before the Native branches emerge.

For context, I've depicted a representative sample of Y chromosome relationships from men around the globe in **Figure 17**. The letter or letter-and-number combinations are simply technical shorthand for branch clusters—or, in technical terms, *haplogroups*—of the tree. Typical European branches are haplogroups *R1a, R1b*, and *I*. Typical African branches are haplogroups *A, B, E2, E1a*, and *E1b1a*. The oldest non-European, non-African branch in the Americas is haplogroup *Q*. It should be visible in **Figure 17** as a branch cluster that is a brother to haplogroups *R1a/R1b/R2* and *S/K/M*.

Globally, haplogroup *Q* is also found in the Old World, especially in Central Asia and northern Eurasia (**Figure 18**). In other words, Old World and New World men belong to the same branch, which implies that they had a common ancestor, which means that they were all once in the same location.

Which location? Closer examination of the details suggests a couple of possible answers.

The deepest branches on haplogroup *Q* broke away from the main trunk and gave rise to people in Eurasia (**Figure 19**). None of the deep branches gave rise to Native Americans. Or did they?

---

7 For reasons unknown, the mitochondrial DNA-based tree is statistically very noisy. It's difficult to nail down dates with any sort of precision. The opposite holds true for the Y chromosome DNA-based tree.

It's theoretically possible that haplogroup Q has been in the Americas for thousands of years. It's possible that people have been migrating *out* of the Americas and *into* Asia for millennia. But let's consider the implications of this scenario.

Let's say that haplogroup Q did indeed originate in the Americas, and that the direction of migration has been primarily *out* of the Americas, not *into* the Americas. Under this scenario, the current data for haplogroup Q imply that migrations happened regularly out of the Americas from about 1000 B.C. to the A.D. 300s to 600s. And then the migrations abruptly, and permanently, stopped.

Interesting. Strange.

Now let's consider the opposite scenario. Let's say that haplogroup Q originated in Central Asia. Under this hypothesis, we still have to explain the abundance of haplogroup Q branches in Asia that broke away from the main trunk of haplogroup Q in the distant past. But, this time, the migrations would be simpler. They would be out of a Central Asian homeland and into the surrounding Asian regions.

This scenario is consistent with the known migrations of Asian history. For example, *Scythians*, long-standing residents of Central Asia, invaded the Middle East in the 600s *B.C.*[8] Lesser-known Central Asian people like the *Yuezhis* invaded India in the centuries around the time of Christ.[9] Central Asians have a long history of interactions with their Asian neighbors to the south.

What about the American side of the equation? This scenario implies that Central Asians left their Asian homelands and arrived in the Americas in the A.D. 300s to 600s. And then, they never left.

Where is the historical record of this migration? Globally, in the first few centuries after Christ, Central Asia was a hotbed of movement. Around A.D. 400, Attila the Hun and his hordes left Central Asia and ripped through the heart of Europe, accelerating the dissolution of the Roman Empire. In the East, the Chinese equivalent of the Roman Empire, the Han Dynasty, fell in the A.D. 200s. In the aftermath, northern

8 Beckwith 2009, p. 61-62.
9 https://www.britannica.com/topic/Yuezhi

China was subsumed by another Central Asian people, the *Xianbei*, who ruled into the A.D. 500s.[10]

Is it reasonable to suppose that, around the same time, another group of Central Asians migrated into America? I think the answer is *yes*. The details of the Y chromosome tree add even more color to this picture. In Europe today, we can find the echo of the fall of the Roman Empire, as seen through the descendants of the invaders. Low levels of haplogroup Q exist among European men. These branches broke away from the main trunk of haplogroup Q around the same time[11] that the Huns were known to have invaded Europe—as if the men of Attila the Hun's armies belonged to haplogroup Q.

What does this mean for the Americas? Let's line up these observations in order and find out: Native American branches belong to haplogroup Q. They arrived in the Americas in the A.D. 300s to 600s. Men from Attila the Hun's armies belonged to haplogroup Q. They also broke away from the main trunk of haplogroup Q around the same time as the American branches did, but they headed in the opposite direction. It would seem, then, that these Native American arrivals were close relatives of the men connected to Attila the Hun.

At first pass, this may seem shocking. But upon reflection, I find this conclusion to be compelling, especially in the big picture context of the history of Central Asia. Historically, the steppes and grasslands of Central Asia have long resisted cultivation and agriculture. Instead, the Central Asian environment has consistently forced its people to adopt pastoral lifestyles. Naturally, herders are always on the move. Consequently, they become skilled at adaptation to new environments. It would be perfectly on-brand for a group of migrating Central Asians to wander far enough that they eventually reached the New World—and thrived.

10 Holcombe 2017, p. 66-71.

11 E.g., see the distribution of *Q-L804* in the FamilyTreeDNA dataset (https://discover.familytreedna.com/y-dna/Q-L804/tree). It splits from the dominantly Native American branches (*Q-Z780* and *Q-M3*) around the same time that these split from one another. According to FamilyTreeDNA, this split was around 13,000 years ago. As per the calculations in Jeanson 2022 on similar datasets, this translates to roughly the A.D. 300s to 600s.

Clues from elsewhere in the Americas reveal just how profound the haplogroup Q arrival was. They also hint at what may have happened in North America after the people of haplogroup Q first appeared.

In modern Latin America, just like in modern North America, the indigenous males belong to European and African Y chromosome tree branches. But not every indigenous male. Once the European and African branches are removed from the discussion, the oldest branch that remains is haplogroup Q. We have no genetic lineage older than haplogroup Q *anywhere* in the Americas.

The historical context for Mesoamerica makes this conundrum more pressing. At the beginning of, or during, the 1st Millennium B.C.,[12] the Maya civilization arose in the jungles of Guatemala or just across the borders in Belize, in southeastern Mexico, and in parts of Honduras and El Salvador (**Figure 20**). They left us their script (**Figure 21**), which is partially hieroglyphic. From these records carved in stone for posterity, we can read the history of the Maya kings and queens. Because they were precise timekeepers and avid observers of the heavens, we know with calendrical precision when many rulers rose and fell. When haplogroup Q arrived in the A.D. 300s to 600s, the ancient Maya civilization was just entering its Classic era.

Contemporary with the Maya, another spectacular polity arose in central Mexico. *Teotihuacan* became a metropolis between 150 B.C. and A.D. 200.[13] Here, too, the Native Americans erected massive pyramids and buildings (**Figure 22**). At its climax, Teotihuacan...

> covered over 8 sq. miles (20 sq. km) and was fully urbanized. Teotihuacan was laid out shortly after the time of Christ on a grid plan that is consistently oriented to 15 degrees 25 minutes east of true north, arguing that the planners must have been sophisticated surveyors as well. Various astronomical explanations have been advanced for this alignment, none of them completely convincing. Perhaps the strangest fact regarding this great city plan is

---

12 Maya civilization seems to have begun to flourish right as the Olmec civilization declined (see Coe and Koontz 2013, p. 61). But early Maya may have been present in Mexico in 1000 B.C. (see Inomata et al. 2020).

13 Coe and Koontz 2013, p. 108.

> that there is absolutely no precedent for it anywhere in the New World.
>
> Teotihuacan's major axis is the Avenue of the Dead, which used to be thought to end at the so-called Ciudadela ("Citadel") in the south, a distance of 2 miles (3.2 km) from its northern terminus at the Pyramid of the Moon. It is now known that the avenue is *twice* this length, and that it is bisected in front of the Ciudadela by an east-west avenue of equal length, so that the city, like the much later Aztec capital, was laid out in quarters.[14]

At its height, it boasted a population of up to 200,000 people.[15]

And yet…the oldest genetic lineage that we have anywhere in the Americas arrived *after* these civilizations were founded. We have no genetic link to these spectacular builders of old.

What happened? For Teotihuacan:

> The city met its end by the end of the sixth century AD through the deliberate destruction of palaces and associated elite art. In one example from the Xalla palace, a rare monumental marble anthropomorphic figure was smashed, the bits scattered, and the entire area burned in what was clearly a desecration of the piece. Similar elite anthropomorphic sculptures, including one found adjacent to the Temple of Quetzalcoatl, were also smashed and scattered. All three of the major palaces…as well as many more along the Avenue of the Dead, were burned to the ground in a targeted conflagration and the art associated with these palaces and their temples was systematically destroyed. … By AD 600 direct Teotihuacan influence over the rest of Mesoamerica had ceased.[16]

14 Coe and Koontz 2013, p. 108.
15 Coe and Koontz 2013, p. 114.
16 Coe and Koontz 2013, p. 123-124.

The Maya fared little better:

> Beginning shortly after AD 750 some Classic centers and polities began to show signs of severe trouble, as reflected in the cessation of dated monuments, abandonment of elite building projects, and changes in ceramic inventories. These symptoms are first evident on the western margins of the Lowlands…Some centers were abruptly abandoned as kings lost their power and the supporting populations disappeared as well. Elsewhere the decline was more gradual, and lesser nobles and some of the supporting population persisted for a time. The last gasps of the old tradition occurred during the Terminal Classic period, and by AD 1000 many great Classic centers, if not entirely abandoned, were no longer occupied by kings and nobles, nor surrounded by large and prosperous populations of rural farmers. What makes this collapse especially puzzling is its apparent abruptness at some centers and the fact that some of the most impressive sculptures and buildings were erected on the very eve of the catastrophe.[17]

Academics, typically reserved in their language, speak of these events in dramatic terms:

> Maya civilization in the Central Area [i.e., the lowlands of Guatemala and adjacent regions] reached its full glory in the early eighth century AD, but it must have contained the seeds of its own destruction, for in the century and a half that followed all its magnificent cities had fallen into decline and were ultimately abandoned. This was surely one of the most profound social and demographic catastrophes in human history.[18]

The big question, naturally, is *why* the civilizations of Teotihuacan and of the Classic Maya dissolved. Internal discontent has been hypothesized.[19] Perhaps the subjects of the Maya rulers grew weary of their kings.

---

17 Webster 2002, p. 46-47.
18 Coe and Houston 2015, p. 174.
19 Webster, D., "The Classic Maya Collapse," p. 329, in: Nichols and Pool 2012.

Or maybe something else was afoot:

> Another form of conflict—intense warfare—was present throughout Maya culture history and peaked during the Late Classic [i.e., right before the Classic Mayan Collapse].[20]

It's very tempting to connect this conflict to the arrival of haplogroup *Q*. I can picture haplogroup *Q* warriors—the relatives of Attila the Hun—marching into Central America and destroying the Maya cities the way Attila destroyed Europe.

But the Maya ruins tell a different story:

> Few [Maya] centers anywhere show signs of violent and widespread destruction.[21]

What do we make of all this? We have the fact that Native American Y chromosome branches go back to only the A.D. 300s to 600s. We also have the fact that some of the biggest Mesoamerican civilizations underwent violent or mysterious collapse in the centuries following the arrival of haplogroup *Q*. If nothing else, this sequence suggests a possible cause-effect relationship.

I can't tell you specifically how the arrival of haplogroup *Q* may have precipitated civilizational collapse. But I can make some reasonable guesses. At Teotihuacan, the *Q* people may have violently overthrown the city. Among the Mayan lowlands, perhaps *Q* invaders brought foreign disease. Or maybe events at Teotihuacan had ripple effects among the Maya polities; archaeology and the records of the Maya suggest that they had some sort of relationship with Teotihuacan.[22] Or maybe the collapse was some combination of the above.

Regardless of the details on how, it's clear that, after the people of haplogroup *Q* appeared, massive political and civilizational change happened in Mesoamerica. Whoever produced the pre-haplogroup-*Q* civilizations disappeared genetically.

◂◂ ▪ ▸▸

20 Webster, D., "The Classic Maya Collapse," p. 329, in: Nichols and Pool 2012.

21 Webster, D., "The Classic Maya Collapse," p. 329, in: Nichols and Pool 2012.

22 Chávez, S.G. and Spence, M.W., "Interaction among the Complex Societies of Classic-Period Mesoamerica," in: Nichols and Pool 2012.

Perhaps the people of haplogroup Q induced a similar catastrophe—albeit on a smaller scale—among the sparse peoples of North America. Whoever spawned the Hopewell culture left no genetic signature of their existence. The haplogroup Q invaders may have wiped out the earlier peoples in one fell swoop. Or they may have slowly replaced them, either by out-reproducing them or occasionally killing off the more successful. Either way, I strongly suspect that the builders of the Effigy mounds were *not* the descendants of the builders of the Hopewell earthworks.

I suppose it's possible that the builders of Hopewell gave rise to the builders of the Effigy mounds, and then eventually were replaced. But the timing of the arrival of haplogroup Q from Central Asia seems to point more toward the Effigy mounds being built by someone new. I can easily imagine a scenario where the initial Q arrivals precipitated the fall of Hopewell. Then, after taking a century or two to get established in North America, and then another century or two to differentiate into more specific tribal groups, one of the tribes commenced construction of a new mound-building culture.

In the A.D. 300s to 600s, North America underwent a dramatic and permanent change—because it was invaded.

# 4

# The Other Event
## A.D. 900s

In the 10th century A.D., something was stirring again in Central Asia. Last time this happened—in the A.D. 300s to 600s—North America was invaded from the west. This time was no exception.

In the A.D. 900s, several little-known groups of Central Asians were on their way or about to head west into Europe. *Magyars, Pechenegs, Oghuz*, and *Kipchaks* are not household names. But these peoples were on the move, just like the Huns were several centuries prior.

Happily, we have much more detail on the A.D. 900s invasion of North America than the one from the A.D. 300s to 600s. It's not just the fact that Central Asians came over. It's that we can connect this event to specific, recognizable Native American nations who survived until and through the Contact era. With this event, we can put names, places, and dates together in a coherent story.

Even better, we can connect this pre-Columbian event to one of the most famous Contact-era encounters in North American history.

◂◂ ▪ ▸▸

In A.D. 1620 when the Pilgrims arrived on the Atlantic shores of North America, they were met by relatives of the Wampanoag. The rest, they say, is history.

But what about the history leading up to this event? We know how the Pilgrims got there. I was never taught how the Wampanoag made their way to Massachusetts.

At first contact, the Pilgrims wouldn't have been taught this history either. The Wampanoag didn't speak English. Communication would have been difficult. Centuries later, scholars would revisit the language of the Wampanoag and find clues. Embedded in their words and syntax were signposts to their story, a story that would once again link the Old World and the New, and rewrite our understanding of this famous first encounter.

Currently, Wampanoag is classified as part of the *Algonquian* subgroup of the *Algic* language family. If the concept of language families is new to you, perhaps an analogy to something familiar will help.

Right now, you're reading or hearing this book in English. But English wasn't always in the form that is currently engaging your senses. If you've ever read Shakespeare, you know that the people of his day spoke and wrote very differently. Shakespeare lived in the A.D. 1500s and 1600s, which means that the English with which I'm communicating wasn't common practice until recently.

Go back deeper in time to the early Middle Ages,[1] and you'll quickly realize that English has changed dramatically. I'm a native English speaker, but I find the poem *Beowulf* impossible to read. The language is foreign to me. Yet it's classified as a work of Old English literature.

At an even deeper period of history, English once was something else entirely. Because my mother is German, I'm familiar with the German language. It's clear to me that there are similarities between German words and English words—similarities which hint at common ancestry. *Mother* in German is *Mutter. Father* is *Vater*, with the first letter pronounced the same way that English speakers vocalize the *f* sound. The parallels go on.

I'm also currently learning Spanish. *Mother* in Spanish is *Madre. Father* is *Padre*. And so on. The similarities between English and Spanish are not as great as between English and German, but they are still recognizable.

1 https://www.britannica.com/topic/Beowulf

Now compare Spanish to Italian. The Italian words for *Father* and *Mother* are identical to the Spanish words—*Padre* and *Madre*. Clearly, Spanish is more similar to Italian than to English or German. Yet none of these four languages is dramatically different from each other. Together, the clues for these four languages suggest a hierarchy. It's not hard to imagine that words for *Mother* and *Father* once existed in a tongue that was ancestral to all four languages. It's also not hard to imagine that this ancestral language, over time, split into the ancestor of English-German and that of Spanish-Italian. And then the English-German ancestral language split into early English and early German, which changed over time into the forms we recognize today. The same process happened for the Spanish-Italian ancestral language.

This sort of analysis has been done for hundreds of languages. Almost 500 languages give clear evidence of descending from the same common ancestor as the English language. Together, they form the Indo-European language *family*. As the name implies, speakers of Indo-European languages reside all the way from India to Iceland (**Figure 23**).

By what mechanism did these languages diversify from the ancestral one? One catalyst for the process seems to be geographic isolation. Regular contact and communication between peoples will keep languages from diverging too quickly. Why start speaking slightly differently if it inhibits your ability to do business with, obtain resources from, and make money off your neighbors? Conversely, if you're self-sufficient and don't engage in contact with your neighbors, then these linguistic constraints disappear. In this case, your language and the language of your neighbors are free to diverge.

At each stage of language comparison, researchers can infer the probable geographic homeland of the ancestral language. Keep going backwards, and eventually you'll arrive at a guess for the original homeland for all the members. You may have noticed that the linguistic descendants of the original Indo-European language live as far apart as Britain and Bombay. For Indo-Europeans, one of the favored guesses for the homelands is north of the Black Sea. From there, the first split sent one group toward Europe and another on a path toward South Asia. Subsequent splits gave rise to the individual language members as we observe them today, geographic separation likely playing a role along the way.

All these principles and processes apply to the Wampanoag language and the Algic language family. At Contact, the members of the Algic family had a profound geographic reach (**Figure 24**). For example, consider some of the more familiar ones. To the west of the Wampanoags, in what is now Massachusetts and New York, the Mohicans resided. South of the Mohicans, in modern New Jersey and Delaware, the Algonquian namesake of the latter state—the Delaware (*Lenni Lenape*)—lived. The Virginia Algonquians of the Powhatan Confederacy planted roots in Virginia.

The Shawnee, who produced one of the best-known Native leaders of the eastern woodlands, Tecumseh, moved about from Ohio down to Cherokee country in the southeast. To the west of the Shawnee were the Illinois. Among the Great Lakes, Potawatomi, Fox, Sac, Kickapoo, Menominee, and Ojibwa/Chippewa were all resident Algonquians. To the north, in the lands of what is now Canada, lived the Cree, Montagnais, Mik'maq, and other Algonquian peoples.

Cheyenne were perhaps the best known of the Algonquian tribes of the Great Plains. Arapaho were near them. Blackfeet, who camped near the Rockies, were one of the western-most Algonquian tribes. The farthest west were the lesser-known Yurok and Wiyot tribes of California, who were the two non-Algonquian nations in the larger Algic family.

Now let's reflect on this image. Again, each of these Algic/Algonquian nations are shown because they are linked—by language. These linguistic links imply a *history*.

◂◂ ▪ ▸▸

Among European languages, we've already witnessed the concept of a language hierarchy. German and English are more related to one another than to Spanish or Italian. Spanish and Italian are more related to each other than to German or English. Yet all four are more related to one another than any are to, say, Bantu languages in Africa.

Similarly, among the Algic language family members, a language hierarchy exists. The deepest linguistic split is between the California languages (Yurok, Wiyot) and the Algonquian languages. Then, among the latter,

> Goddard...presents Algonquian as a west-to-east cline, not of genetic subgroups but of chronological layers, with the greatest time depth found in the west and the shallowest in the east. That is, each layer, in his view, is distinguished from those to the west by innovations and from those to the east by archaic retentions, where each wave of innovations is farther to the east, giving the characteristic clinal configuration that reflects the general west-to-east movement of the family. Blackfoot (in the West) is the most divergent; Arapaho-Atsina [i.e., Gros Ventre] and Cree-Montagnais are the second oldest layer. The next oldest includes Arapaho-Atsina [Gros Ventre], Cree-Montagnais, Cheyenne, and Menominee; next is Core Central Algonquian languages; the final layer is Eastern Algonquian (the only grouping or layer that constitutes a valid subgroup).[2]

A couple of clarifications: This paragraph repeats *Arapaho-Atsina* and *Cree-Montagnais*. For sake of simplicity, I'm going to treat them as belonging strictly to the second oldest layer, with the next oldest constituting just Cheyenne and Menominee. Also, from the classification table given in the book I quoted above, the Core Central Algonquian languages include Fox, Sac, Kickapoo, Mascouten, Shawnee, Miami-Illinois, Ojibwa-Potawatomi-Ottawa, Algonquin, and Saulteaux. Finally, the Eastern Algonquian languages are easily visualized in **Figure 25**. They include the Atlantic coast Algonquians, from the Mik'maq north of Maine to the Algonquians in North Carolina. The Eastern Algonquian languages also include Wampanoag.

This language hierarchy intimates a sequential origin for the members of this language family. The deepest Algic linguistic split is between the westernmost nations (Yurok, Wiyot) and the Algonquians. This implies that the Yurok and Wiyot split off first from the main body of Algic speakers. Among the latter, the Algonquian language hierarchy proceeds from west to east (see **Figure 25, see also Figure 24**). This

2 Campbell 1997, p. 152-153.

implies that the next-earliest splits were in the west; the last, in the east, with one of the last steps being the separation of the individual members of the Eastern Algonquians—including Wampanoag—from one another into the individual nations present at Contact.

Apparently, the Pilgrims met one of the most recent Algic groups to form.

◂◂ ▪ ▸▸

Where exactly did the ancestor of the Wampanoags—and of the Algics in general—live? At a minimum, the homelands of the Algics would seem to have been somewhere in the far western areas of the modern United States or Canada, somewhere between where the Yurok, Wiyot, and Blackfeet were encountered at Contact (**Figure 25**). This region would be consistent with the language hierarchy that we just discussed, with the west-to-east pattern of linguistic progression.

But the language hierarchy can take us only so far. Implicit in what I just proposed is an assumption—that we need to keep the reconstruction as simple as possible, that we need to minimize the amount of distance each group migrates. I didn't assume that we need to keep the reconstruction close to historical reality.

It's entirely possible that the Algic ancestors moved all over the map before the first language splits happened. In fact, the map of the Algics at Contact shows that a tremendous amount of migration occurred once language splits started happening (**Figure 25**). At Contact, Algics were found from California to northeast of Maine. Let's say that the original ancestors of the Algic lived somewhere in the vicinity of modern Washington state, or even Oregon or British Columbia. To explain the coast-to-coast distribution at Contact, the descendants of the ancestral Algic peoples, including the Wampanoag, would still have had to migrate clear across the entirety of North America, all the way to the Atlantic Seaboard. If this much migration happened even when we try to minimize it, imagine how much may have happened without such artificial constraints.

◂◂ ▪ ▸▸

In **chapter 1**, I referenced the *Red Record*, a document that details the history of the Delaware people, the *Lenni Lenape*. At first pass it seems to be an account of just the Delaware. Deeper examination revealed something else—something intimately linked to the history of the Wampanoags and to the language comparisons that we just made.

The *Red Record* describes a long west-to-east migration of the Lenni Lenape. In 1993, David McCutchen synthesized all the geographic hints in the *Red Record* and mapped out the path of their migration.[3] **Figure 26** is a depiction of what he derived.

Do you notice something unusual about **Figure 26**? Can you see the similarities between McCutchen's path and the general path of migration that we derived from linguistics (**Figures 24-25**)? This initial observation suggests that the history embedded in Algic linguistics and the history in the *Red Record* tell the same story.

This conclusion grows stronger in light of another aspect of the *Red Record.* Six times it explicitly mentions when a subgroup of the main body of people migrated away. Contextual clues imply a total of seven (see **Appendix B** for documentation).

Do you recall the Algic language hierarchy? The first and oldest division separated Yurok and Wiyot from the Algonquian languages. Recall as well that language division tends to imply geographic—and population—separation. The next oldest division separated Blackfoot from the remaining Algonquian languages. Then the Arapaho-Atsina [i.e., Gros Ventre] and Cree-Montagnais split off. The fourth split moved Cheyenne and Menominee away.

Among the remaining Algonquian languages are the Core Central Algonquian languages—the Fox, Sac, Kickapoo, Mascouten, Shawnee, Miami-Illinois, Ojibwa-Potawatomi-Ottawa, Algonquin (Algonkin), and Saulteaux languages. The Core Central languages represent subdivision five. The only remaining languages are the entire group known as the Eastern Algonquian, which includes, as we observed earlier, the Mik'maq north of Maine to the Algonquians in North Carolina (**Figures 24-25**). At some point, the Eastern Algonquian languages all separated from one another; this event represents the sixth split in this sequence.

3 For details, see McCutchen 1993.

In the *Red Record*, the seventh and final population split is the separation of the various Delaware tribes from one another—into the Unami, Munsee, and Turkey tribes (see **Appendix B**). All three languages are part of the Eastern Algonquian subgroup of Algic languages. The sixth split (**Appendix B**) happens when "The eastern people, the Wolf People, moved east and north"[4] relative to the main body that was in the vicinity of Niagara Falls (see **Figure 26**). Today, the Mohegans call themselves the "Wolf People."[5] Like the Delaware language, the Mohegan language is also part of the Eastern Algonquian subgroup. In other words, the sixth and seventh splits in the *Red Record* would not be linguistically distinguishable by the categories we've been discussing.

But the first five splits, along with the combined last two, add up to a total of six.

The six Algic linguistic separations find an echo in the first six recorded population separations in the *Red Record*.

It would seem, then, that the *Red Record* documents the migration, not just of the Delaware, but also of the entire Algic language family—including the ancestors of the Wampanoag.

◀◀ ▪ ▶▶

The *Red Record* migration doesn't begin in Washington state, or in Oregon, or in British Columbia. It begins in Asia and leads across what we now call the Bering Strait:[6]

> After the Flood, the Lenape, the True Men, the Turtle People, were crowded together, living there in cave shelters.
>
> Their home was icy. Their home was snowy. Their home was windy. Their home was freezing. …
>
> By the dark fish sea, the gaping hollow sea, settled the White Eagle clan and the White Wolf clan.
>
> Rowing, crossing the water, for long they gloried in the eastern light, in the land of Akomen.

4 McCutchen 1993, p. 124.
5 https://www.mohegan.nsn.us/about/our-tribal-history
6 I'm largely following McCutchen's 1993 geographic reconstruction. For details, see McCutchen 1993.

> Beaver Head and Great Bird, they said "Let us go!" "To Akomen!" they said. …
>
> On a wondrous sheet of ice all crossed the frozen sea at low tide in the narrows of the ocean.
>
> Ten times a thousand they crossed; all went forth in a night; they crossed to Akomen, the East Land, crossing, marching, marching everyone together.[7]

For the Wampanoag and their ancestors, Massachusetts was simply the last stop on an incredibly long west to east journey—one that began on an entirely different continent.

◂◂ ▪ ▸▸

When did the ancestors of the Wampanoag, and of the entire Algic language family, migrate from the Old World into Alaska? From **chapter 3**, we already have a starting place for a good guess: We know that haplogroup Q arrived in the Americas in the A.D. 300s to 600s. It's entirely possible that the Algics were part of the haplogroup Q migration.

Were they? The *Red Record* does not state a calendar date for when the Algics first arrived in Akomen. But it does contain stanzas to which we can assign explicit time stamps. These time stamps allow us to reason backwards in time to estimate an arrival time.

The first time stamp comes via the translator McCutchen. He references[8] Eli Lily,[9] who references an A.D. 1768 document[10] in which the author relates the existence of Delaware wampum records. These records trace back to the time when the Delaware first arrived at the Atlantic. According to the wampum, the year was A.D. 1396.

In the *Red Record*, we find a description of this event, as well as the name of a specific leader or *sachem* who ruled at that time:

> All the Hunters reached the Sun's Salt Sea; once more, the Ocean.

---

7 McCutchen 1993, p. 68, 72, 76.

8 McCutchen 1993, p. 18.

9 Lily, Eli, 1954. "Speculations on the Chronology of the Walam Olum and Migration of the Lenape," in: N.A. 1954.

10 Beatty 1768.

> Red Arrow was the sachem at the tidewater.[11]

The second time stamp occurs at the end of the *Red Record*. The book closes with the Delaware on the Atlantic shores, looking east:

> White Crab was the sachem, friend of the shore.
>
> Watching closely was the sachem, looking seaward.
>
> For at that time from north and south, the white people came.
>
> Friendly people, in great ships; who are they?[12]

McCutchen connects this event to the arrival of Europeans in A.D. 1620.

The two time points allow us to estimate the length of time each sachem ruled.[13] Let's say that the final sachem, *Watching Closely*, ruled when the Europeans arrived in A.D. 1620. *Red Arrow* was the sachem when the Delaware first arrived on the Atlantic in A.D. 1396. After *Red Arrow*, the *Red Record* lists around 22 sachems,[14] ending with *Watching Closely*. If we divide the number of sachems into the difference in calendar years,[15] we can estimate 9.8 years per sachem.[16]

The first sachem who led them from Asia into Akomen was *White Eagle*. After *White Eagle*, the *Red Record* lists around 96 sachems,[17] ending with *Watching Closely*. If we multiply the number of sachems by the years-per-sachem in the previous paragraph,[18] the first arrival in Akomen appears to have been in the range of A.D. 522 to A.D. 746. This overlaps the timing of the haplogroup Q arrival in the New World

---

11 McCutchen 1993, p. 124.

12 McCutchen 1993, p. 136.

13 In theory, we can count the exact number of sachems between the first arrival of the Delaware, and the close of the *Red Record*. But there are two stanzas missing near the end of the *Red Record*. Based on the pattern in prior stanzas, this gap might represent the rule of anywhere from zero to four sachems. So we're stuck with an estimate (see **Supplemental Table A** <answersingenesis.org/go/theyhadnames/> for more details).

14 Range is 20 to 24 sachems.

15 1620 - 1396 = 224

16 i.e., the average of 9.3 and 11.2

17 Range is 94 to 98 sachems.

18 i.e., 9.3 and 11.2
Thus: A.D. 1620 - (9.3 yrs per sachem * 94 sachems) = A.D. 746
A.D. 1620 - (11.2 yrs per sachem * 98 sachems) = A.D. 522

(i.e., A.D. 300s to 600s). It's not a perfect overlap, but it's enough to connect the Algonquians to the haplogroup Q migration.[19]

We can also estimate a second, different date for the arrival of the Algonquians in the New World. The *Red Record* contains another event which can be time-stamped:

> Mistaken was the sachem about what then came.
>
> For at this time, from the Dawn Sea, the Whites appeared.[20]

McCutchen[21] connects this sighting to the known voyages of the Italian explorer Giovanni da Verrazzano along the Atlantic seaboard. da Verrazzano was known to have passed the Atlantic Seaboard in A.D. 1524.

Let's repeat our calculations, but with this new time stamp in view. After *Mistaken*, the *Red Record* lists 13 sachems, ending with *Watching Closely*.[22] From A.D. 1524 to A.D. 1620, 96 years elapsed. If we divide 96 by 13, we can estimate that each sachem led for approximately 7.4 years.

Now apply this number to the entirety of the Algic time in the New World. After *White Eagle*, the *Red Record* lists about 96 sachems,[23] ending with *Watching Closely*. At 7.4 years per sachem, we can estimate the arrival in Alaska to be between A.D. 896 and A.D. 926.

This is several centuries after haplogroup Q arrived in the New World.

Which estimate is correct? A combination of population numbers and modern genetics held the answer.

In 2016, the census of Canada reported population numbers for the First Nations still living in Canada. Algics represented 72% of

---

19 For genetic reasons that are shown in the paragraphs that follow, I suspect that the original author's claim about the Delaware wampum records was wrong. I wonder if the A.D. 1396 date for the arrival on the Atlantic Ocean was true, not for the Delaware, but for the Ojibwa, who I discuss in a later chapter.

20 McCutchen 1993, p. 128.

21 McCutchen 1993, p. 19.

22 The gap in the *Red Record* precedes this section; hence, no range of numbers.

23 Range is 94 to 98 sachems.

the Canadian First Nations population.[24] In contrast, to the south of Canada in the United States, the 2010 census reported that the Algics represented only 15% of the Native American population total.[25]

In 2022, FamilyTreeDNA (FTDNA), one of the major genetic testing companies, released their database of the results of their 200,000+ male Y chromosome testers. FTDNA did not report the Native American nation or First Nation affiliation for each of their testers. Instead, they simply identified the country of residence—e.g., United States, Canada, Mexico, etc.

Included in the FTDNA database were results for haplogroup *Q*. The database also included results for another branch found in Native Americans and Central Asians, haplogroup *C*. Today, haplogroup *C* can be found on several continents, but especially in northern Eurasia (**Figure 27**). In the Americas, it tends to be found at its highest levels in North America (**Figure 27**).

The comparative FTDNA (Y chromosome) data for the United States and Canada were as skewed as the census percentages for the Algic populations. Among U.S. residents, haplogroup *C* represented just 6% of the total population of testers residing on Native American branches. The remainder of the testers (94%) belonged to haplogroup *Q*. In contrast, among Canadian residents, haplogroup *Q* represented 4% of the total, and haplogroup C, 96% (see **Supplemental Table D** online for details).

Clearly, haplogroup *C* roughly correlated with the percentage of Algic populations in both countries. It would seem, then, that the Algic language family arose in the Americas from haplogroup *C*, not haplogroup *Q*.

---

24 Statistics Canada, 2016 Census of Population, Statistics Canada Catalogue no. 98-400-X2016167, https://www12.statcan.gc.ca/census-recensement/2016/dp-pd/dt-td/Rp-eng.cfm?TABID=2&LANG=E&A=R&APATH=3&DETAIL=0&DIM=0&FL=A&FREE=0&GC=01&GL=-1&GID=1334853&GK=1&GRP=1&O=D&PID=110522&PRID=10&PTYPE=109445&S=0&SHOWALL=0&SUB=0&Temporal=2017&THEME=122&VID=0&VNAMEE=&VNAMEF=&D1=0&D2=0&D3=0&D4=0&D5=0&D6=0, accessed January 13, 2023. See **Supplemental Table B** <answersingenesis.org/go/theyhadnames/> **and Appendix B for details of the calculations.**

25 2010 Census of Population and Housing, American Indian and Alaska Native, Summary File: Technical Documentation, U.S. Census Bureau, 2012. On October 10, 2019, I downloaded Table 1. American Indian and Alaska Native Population by Tribe for the United States: 2010 (Internet release date: December 2013). See **Supplemental Table C** <answersingenesis.org/go/theyhadnames/> for details on the calculations.

From the global Y chromosome family tree, we have little published information on the Native American haplogroup *C* branches. However, from the little information that we do have, we can estimate dates for when the branches split from one another. On the haplogroup *C* branch, the closest family tree relatives to American haplogroup *C* are the north/northeast Asian haplogroup *C* males. These two populations separated in the A.D. 800s to 1000s.[26]

This fits the later date (A.D. 896 and A.D. 926) from the *Red Record* for the arrival of the Algics in the Americas. Thus, based on genetics and on their own history, the Algics arrived in Alaska from Asia around the A.D. 900s.

◀◀ ▪ ▶▶

How long had the Wampanoag been in Massachusetts before the Pilgrims landed in A.D. 1620? We now have the tools to make a pretty good guess.

Linguistically, the Wampanoag are part of the Eastern Algonquian subgroup of languages. The *Red Record* doesn't explicitly identify this language subcategory. But it does identify the time of origin for one of the subcategory members—the *Wolf People*, or *Mohegans*. The timing of this event represents the upper boundary for when the Wampanoag formed as a people. In other words, if we could date the time of origin of the Mohegans, then the Wampanoag likely formed around that time, or more recently in history.

We can estimate a date thanks to the derivation in the previous section. Again, the *Red Record* has a long chain of sachems stretching from the first arrival of the Algics to the formation of the Mohegans. Now that we know the approximate year for first arrival in the Americas, we can estimate the date for the rule of all the subsequent sachems. For the Mohegan breakaway event, which occurred when *Good Inscribed* was the sachem, the approximate year is between A.D. 1450 and A.D. 1480.[27]

---

26 See **Supplemental Table E** <answersingenesis.org/go/theyhadnames/> for details of the calculations.

27 See **Supplemental Table A** <answersingenesis.org/go/theyhadnames/> .

In other words, just a century and a half before the Pilgrims reached Cape Cod, the Wampanoags arrived in Massachusetts.[28]

◂◂ ▪ ▸▸

The Algic migration to the Americas was the last before European arrival—at least, the last detectable one. The first migration[29] spawned the Poverty Point and Hopewell cultures. These early peoples were then replaced by the invasion of the peoples belonging to haplogroup *Q*. The Algic invasion (haplogroup *C*) brings the total number of migrations to three.

At this juncture, our ability to reconstruct events increases dramatically. From A.D. 900 onwards, the indigenous histories have much to reveal. This era takes us through one of the greatest empires and battles in all of pre-European North American history (**chapters 6 and 7**), as well as the drama that led up to it (**chapter 5**) and followed it (**chapter 8**). For one specific region of North America, the southwest, the centuries following A.D. 900 produced the most intense periods and most memorable set of ruins in the entire continent (**chapter 9**). In other words, the centuries before Contact filled in the map of pre-European North America with more detail than we've ever seen before.

28 There is a caveat to this statement. When the *Red Record* describes the origin of the Shawnee, it says that the Nanticokes arose at the same time (see **Appendix B**). Linguistically, the Shawnee belong to the Core Central Algonquian language subgroup; the Nanticokes, to the Eastern Algonquian language subgroup. Based on the linguistic discussion from an earlier section of this chapter, it would seem that the main bulk of the Eastern Algonquians must have arisen *after* the formation of the Core Central Algonquians. If they arose simultaneously, then they should all form a single large subgroup, not two different ones. Nonetheless, it is formally possible that the Wampanoags arose at the same time as the Shawnee, which would bump their origins back by about a century: The estimate date range is A.D. 1362 to A.D. 1391 (see **Supplemental Table A** <answersingenesis.org/go/theyhadnames/> for justification for the Shawnee origin date range). Even this early date range is only two and a half centuries before the Pilgrims reached Cape Cod.

29 Or sets of migrations.

# Part II

# Centuries Before Contact

# 5

## Before the Crisis
## A.D. 900s to 1250

Seven hundred years before Europeans lumbered up the shores of the Atlantic Coast in New England, the Algic lumbered up the Pacific shores of the land of the midnight sun. Several centuries earlier, another group of Central Asians had landed on these same Arctic shores. Now, marching down to meet the new Algic arrivals were the—well—who? Who was in Alaska in A.D. 900? What did the map of North America look like half a millennium before Europeans would permanently set foot in the New World? The biggest clues to these questions come from unusual sources.

At Contact, the Algics were one of over 60 language groupings in North America (see online map.)[1] Their story is just one of many. In this chapter, we'll follow their drama through the lens of the *Red Record*. However, as we do so, we'll end up learning the stories of so many more nations who won and lost, ruled and fled in the vast landscapes of this continent.

◂◂ ▪ ▸▸

Almost as soon as the Algic reached the Arctic coast of the New World, the battles began:

1 <answersingenesis.org/go/theyhadnames/>

> All declared, "Kolawil, Noble Elder, thou are sachem here [in Akomen]."
>
> "The Snakes come; thou must slay them; the Snake hollow they must leave."
>
> Defeated, all the Snakes fled to hide in the swampy vales.
>
> After Kolawil in the evergreen land the sachem was the White Owl.
>
> After him, the sachem was Constantly on Guard.[2]

Who were the *Snake*s that the Algics overcame in the early A.D. 900s? Who did *Constantly on Guard* have to guard against? At Contact, members of the *Eskimo-Aleut* language family covered the entire western coast of Alaska (**Figure 28**). If they had covered the same area in the A.D. 900s, they would be a strong candidate for the *Snakes* that the Algics first encountered.

What was their location in the 10th century? Genetics provides a clue.

Today, nearly all Native American members of haplogroup *Q* belong to one of two specific subbranches on the DNA-based tree. The names aren't pretty, but they follow current academic practice. The subbranches are known as *Q-M3* and *Q-Z780*. The Eskimo-Aleut also can be found on a third subbranch known as *Q-NWT01*.[3]

In terms of timing, the Siberian/Arctic *Q-NWT01* branch is like the other haplogroup Q subbranches. All seem to have split from the rest of the Asian branches around the A.D. 300s to 600s.[4] In other words, Eskimo-Aleut would have been present in the Americas when the Algics showed up several centuries later.

Where in the Americas? Current archaeological thinking proposes a migration of Eskimo-Aleuts out of the Alaska area toward Greenland. The timing of the *Thule* journey is right around the time that

---

2 McCutchen 1993, p. 82, 86.
3 Dulik et al. 2012; Olofsson et al. 2015; Luis et al. 2023.
4 See technical justification in **Appendix B**.

Algics landed.[5] Makes me wonder if the migration was prompted by conflict with the Algics—and the victory of the latter.

The early Algic arrivals were a strong and martial people.

◂◂ ▪ ▸▸

The Eskimo-Aleut weren't the only potential rivals for the newly-arrived Algics. At Contact, northwestern Canada and most of Alaska were dominated by members of the *Athabaskan* language family (**Figure 29**).

Like the Algics, the Athabaskans trace their origins to Asia. One of the Athabaskan language family members, the *Chipewyans* (or *Chepeweyans*), residents of north-central Canada at Contact (see **Figure 29**), said so:

> In the voyages of Sir Alexander Mackenzie among the Arctic tribes, he relates of the Chepeweyans, that "they have a tradition that they originally came from another country, inhabited by very wicked people, and had traversed a great lake, which was narrow and shallow, and full of islands, where they had suffered great misery, it being always winter, with ice and deep snow. Their progress (the great Athapasca family) is easterly, and according to their own tradition, they came from Siberia."[6]

The critical question, of course, is when this migration occurred.

The answer lies in a data comparison. It's similar to the one we used in **chapter 4**. Let's compare the census numbers of Athabaskans to the frequency of genetic branches.

---

5 Mainstream science connects the Thule to the Eskimo-Aleut at Contact (e.g., see Pauketat and Sassman 2020, p. 280). The timing of the Thule migration is debated. It may have begun in the A.D. 900s—exactly when the Algic arrived in Alaska (Pauketat and Sassman 2020, p. 294-296). Or it might have commenced a century later (e.g., see Fagan 2019, p. 109). Either way, it's hard not to link the two groups in a cause-effect manner.

6 Schoolcraft 1851, p. 19. I retained his original spellings.

In Canada by census,[7] Athabaskans represent just 4.8% of the total First Nation population.[8] In the FamilyTreeDNA (FTDNA) dataset that we discussed in **chapter 4**, the indigenous branches of the family tree (i.e., haplogroups *C* and *Q*) contain only 23 total Canadian males.[9] If we multiply 4.8% by 23 males, we get 1 male. In a perfect world, we would (barely) be able to detect the Canadian Athabaskans in the FTDNA dataset. In the real world of statistical variability, it would be no surprise if the genetic signal of the Canadian Athabaskans was undetectable.

In the United States, Athabaskans are one of the largest Native American populations.[10] At 21%, their total is matched only by the members of the Iroquoian language family (18% of the total). The latter percentage is due in large part to the number of a specific branch of the Iroquoian family, the Cherokee. In the FTDNA dataset, the indigenous branches of the family tree contain 109 U.S. males. If we multiply 21% by 109 males, we get 23 males. In other words, a U.S. Athabaskan branch should be easily detectable.

The *Q-M3* subbranch of haplogroup *Q* has its own subbranches. One of them, *Q-Y4300*, is found at an intriguing frequency. Of the total U.S.-based testers in haplogroups *C* and *Q*, about 22% belong to *Q-Y4300*.[11]

By itself, this result could be either Athabaskan or Iroquoian. Granted, it's a near-perfect match to the population frequency of Athabaskans. But 22% is close to 18%. It could be Iroquoian.

---

7 Statistics Canada, 2016 Census of Population, Statistics Canada Catalogue no. 98-400-X2016167. https://www12.statcan.gc.ca/census-recensement/2016/dp-pd/dt-td/Rp-eng.cfm?TABID=2&LANG=E&A=R&APATH=3&DETAIL=0&DIM=0&FL=A&FREE=0&GC=01&GL=-1&GID=1334853&GK=1&GRP=1&O=D&PID=110522&PRID=10&PTYPE=109445&S=0&SHOWALL=0&SUB=0&Temporal=2017&THEME=122&VID=0&VNAMEE=&VNAMEF=&D1=0&D2=0&D3=0&D4=0&D5=0&D6=0, accessed January 13, 2023.

8 See **Supplemental Table B** <answersingenesis.org/go/theyhadnames/> for details on the calculations.

9 See **Supplemental Table D** <answersingenesis.org/go/theyhadnames/>.

10 2010 Census of Population and Housing, American Indian and Alaska Native, Summary File: Technical Documentation, U.S. Census Bureau, 2012. On October 10, 2019, I downloaded Table 1. American Indian and Alaska Native Population by Tribe for the United States: 2010 (Internet release date: December 2013). See **Supplemental Table C** <answersingenesis.org/go/theyhadnames/> for details on the calculations.

11 See **Supplemental Table D** <answersingenesis.org/go/theyhadnames/>.

Additional genetic results resolve this ambiguity. Some of the sub-branches from *Q-Y4300* extend to living men in Mexico.[12] At Contact, it was the Athabaskans (e.g., Navajo and Apache; see **Figure 29**), not the Iroquoians (**Figure 30**), who lived along the modern northern Mexican border. Also, one of the Navajo Code Talkers, of World War II fame, gave a sample of his DNA before he passed. His branch is part of *Q-Y4300.*[13]

Athabaskans trace their origins along haplogroup *Q-Y4300.*

Now we can line up these data to infer a conclusion about the Algic and their enemies: Haplogroup *Q* arrived in the A.D. 300s to 600s. Athabaskans likely belong to the haplogroup *Q* subbranch known as *Q-Y4300.* Therefore, Athabaskans landed in the Americas several centuries before the Algics arrived.

But were the Athabaskans located in the relevant regions? Were they in northwest Canada and Alaska at that time? Or were they much farther away, in the southwest of what is now the United States, before migrating north at a much later date?

Language clues lead the way to the answer.

Up to this point, I've been using shorthand to describe the Athabaskans. Technically, linguists refer to this family as the *Eyak*-Athabaskan family. The deepest linguistic split is between the Eyak language and the Athabaskan subgroup of languages. At Contact, Eyak was spoken in Alaska; the Athabaskan languages, many other places (**Figure 31**). This would seem consistent with an early residence in the far northwest of North America.

Mescalero Apache history confirms this. At Contact, the Mescaleros resided along the modern U.S.-Mexico border (**Figure 29**). Their own history points to an origin much farther north.

The evidence derives from the details of the Mescalero girls' puberty ceremony. The four-night event recapitulates the tribe's history of migrations and contains clues to their distant origins:

---

12 See *Q-Y4300* at https://discover.familytreedna.com/y-dna/Q-Y4300/tree (I accessed on August 2, 2024).

13 Ibid.

> The Mescalero were made in a Land of Ever Winter, near a lake you could not see across … Great Slave Lake or Lake Athabasca, according to different variations of the story. … The northern location of this Apache ancestral land is demonstrated and remembered by references to people, places, and natural phenomena of the region. … It so happened that while the Mescalero were in these northern regions, troubles developed between them and other peoples, some of them relatives. There was a great catastrophe, and the Mescalero were being devastated; they were fighting near the shore of Lake Athabasca. … Following this battle, the Apache were surrounded by goodness and beauty; they became a people and started drifting south.[14]

The *Snakes* who greeted the Algics upon their arrival in the Americas, and whom the Algics defeated, could very well have been Eyak-Athabaskans.

◂◂ ▪ ▸▸

After the Algic victory over their initial enemies, peace followed for a few years and the population expanded.[15]

Perhaps because of this growth, the first split—the likely origin of the Yurok and Wiyot peoples—occurred. It was during the rule of the sachem named *Chilili* (estimated date range of A.D. 926 to 955; see also **Appendix B**).

At this stage in the journey, McCutchen's geographic reconstruction (**Figure 26**) puts the main body of Algics northwest of what is now British Columbia, Canada, in the southeast of what is now Alaska. This is where the population split likely occurred.

This is also nowhere near the geographic position of the Yurok and Wiyot at the time of Contact (see **Figure 24**).

Perhaps the Yurok and Wiyot changed course after the events that came next:

---

14 Carmichael, David L. and Farrer, Claire R., "We do not forget; we remember: Mescalero Apache origins and migration as reflected in place names," p. 187 in: Seymour 2012.

15 McCutchen 1993, p. 86.

> After Chilili the sachem was Ayamek, the Seizer; he slew them all:
>
> The Charlatans, the Snakes, the Blackened Ones, the Stoney Ones.
>
> After Ayamek, ten sachems; much evil was then south and eastward.[16]

Everywhere else in the account of the Algic's New World migrations, the *Red Record* gives the names for the sachems. But in the stanza above, which happened somewhere in the A.D. 930s to early A.D. 1000s, ten sachems pass without a single mention of their identity. It makes me wonder what happened and who was the cause. Was the evil perpetrated by the Algics themselves? Was the lack of names due to shame over what transpired? Perhaps the Yurok and Wiyot fled south to escape their violent relatives.

During this evil time, who did the Algics engage? As per McCutchen's reconstruction, the Algics were in western/southwestern British Columbia during this time (**Figure 26**). At Contact, this was, or was near, Athabaskan country (**Figure 29**). Perhaps earlier in the *Red Record* the sachem *Constantly on Guard* received his name due to ongoing threats from the Athabaskans—threats which were now realized.

At the time of Contact, this region of the west coast was also near members of the *Salish* language family (**Figure 32**). Naturally, it would be tempting to hypothesize the Salish as participants in the "much evil." But the west was not the first home of the Salish. They migrated from elsewhere, according to the *Thompson* branch of the Salish:

> A long time ago the people were living in another country, near a large lake, where they were attacked by enemies. Since they could not cross the water, they were hemmed in by their enemies, and were in danger of being destroyed. Their two chiefs called the men together for a council and dance. ... That night ice covered the lake, and at daybreak the tribe crossed to the other side. ... Yet the people thought, 'Even here our enemies may follow us.' Therefore

16 McCutchen 1993, p. 86, 90.

> they travelled still farther away. They camped for a time, but the location was unsuitable. Four times they moved, for they were dissatisfied with their camps. One place had an insufficiency of wood, another of fish, another of game. At last they came to a country where wood, bark, fish, and game were plentiful, and there they remained. This place, it is said, was the Thompson country at Lytton. They became the ancestors of the Thompson Indians [see **Figure 32**]. On their journey they crossed a great lake, a great plain, a great forest, and a great mountain.[17]

From which direction did the Thompson arrive?

> In the beginning, our ancestors lived at some place inland, south, or southeast from here, on the far side of a great body of fresh water. This was their original home.[18]

These geographic details provide clues as to the likely place of origin for these migrations. It also brings us closer to knowing the relevance of the Salish to the time of "much evil."

The last of their four stops in the journey was "a great mountain." At Contact, the Salish—and in particular the Thompson branch of the Salish—were located on the Interior Plateau of what is now British Columbia. They were bounded on the west by the Pacific Ranges and on the east by the Rocky Mountains. Surely they would have crossed *a great mountain* to reach the Interior Plateau.

The second-to-last step in their journey was a trip across "a great forest." Immediately prior, they traveled across "a great plain." Southeast of their Contact-era position, the "Northwestern Forested Mountains" ecoregion[19] hugs the Rockies (**Figure 33**). Southeast of the "Northwestern Forested Mountains" ecoregion is the "Great Plains" ecoregion[20] (**Figure 33**). This seems to fit the direction and order of the ancestral Thompson migration path.

---

17 Teit, Farrand, Gould, and Spinden 1917, p. 48-49.
18 Teit, Farrand, Gould, and Spinden 1917, p. 51.
19 https://gaftp.epa.gov/EPADataCommons/ORD/Ecoregions/cec_na/NA_LEVEL_I.pdf https://www.epa.gov/eco-research/ecoregions-north-america
20 https://gaftp.epa.gov/EPADataCommons/ORD/Ecoregions/cec_na/NA_LEVEL_I.pdf https://www.epa.gov/eco-research/ecoregions-north-america

The "Great Plains" ecoregion just touches the western edge of Minnesota before giving way to forest (**Figure 33**). Minnesota is *the land of 10,000 lakes*. If the first step on the ancestral Thompson journey was crossing "a great lake," I can easily see Minnesota as the ancestral homeland.[21]

On a map, the length of the ancestral Thompson journeys becomes clear (**Figure 34**).

The biggest question, though, is *when* the Salish arrived. I haven't found an account that gives a time stamp. I also haven't yet identified the Salish Y chromosome lineage. Nevertheless, archeology provides a clue.

The relevant area is the *Northwest Coast*, the geographic region roughly from southern Alaska to coastal Oregon, which overlaps with Contact-era Salish regions (**Figure 32**).[22] The last temporal division in this area is the *Late Pacific Period*, dating from A.D. 200/500 to the late A.D. 1700s. Here, the evidence suggests that Salish, if not other Northwest Coast peoples, preceded the Algics:

> Most archaeologists working on the coast feel that the cultures of the Late Pacific differed little, if at all, from those observed and recorded by the first European visitors to the coast. In this view, the people of the Late Pacific are both the direct biological and cultural ancestors to the coast's modern Native peoples.[23]

Recall that the Algic seem to have been the losers in the "much evil" conflict. Perhaps the Salish were residents of British Columbia before the Algic arrived, stood their ground in the face of these new foes, and then remained there after winning.

◀◀ ▪ ▶▶

The "much evil" of this period may have involved yet another category of peoples. The conflict may have arisen from nations who were

---

21 Given winter ice cover on the Great Lakes (https://www.epa.gov/climate-indicators/climate-change-indicators-great-lakes-ice-cover), I suppose it is also possible that the Thompson crossed one of the Great Lakes.
Special thanks to Matthew Cserhati for the heads-up on this point.

22 Ames and Maschner 1999, especially p. 66, 94-95.

23 Ames and Maschner 1999, p. 95.

*not* resident in western/southwestern British Columbia at Contact. This latter solution might solve two puzzles at once.

The first puzzle is the identity of the enemies. The second puzzle is one I haven't mentioned explicitly yet. You may have noticed it when you first looked at the Contact map of North America. Do you see the unusual linguistic makeup of the western United States (**Figure 35; see also online map**[24])? Do you see how Balkanized the region was? Do you see how many small, individual language families existed in close geographic proximity? And do you notice how this geographic distribution is distinct from the pattern across the rest of the United States? Only the Gulf Coast comes close to this particularization of peoples.[25]

How did this pattern come about? Two of the groups on the west coast came from language families we've already discussed–the Yurok and Wiyot. Several small Athabaskan groups were their neighbors (**Figure 35**). I've already wondered if the "much evil" sent the Yurok and Wiyot south and west to California. I wonder if the "much evil" involved battles with the Athabaskans, such that some of the Athabaskans also went south and west.

But the Balkanized west coast of the United States involves much more than Algic and Athabaskan groups. It includes a litany of peoples you've probably never heard of: Yokutsans, Chumashans, Salinans, Utians, Maiduans, Wintuans, Yukians, Pomoans, Shastans, Takelmans, and many others who lived there at Contact (**Figure 35**). Perhaps the "much evil" involved conflict with these nations who, at the time of Contact, no longer called British Columbia home—because they had been forced to flee south and west.

◂◂ ▪ ▸▸

Thankfully, the "much evil" was followed by an extended time of peace (from around A.D. 1015 to A.D. 1110).[26] The *Red Record* makes no mention of wars or conflicts during this span. The names of the sachems also reflect the stability of this period: *Peaceful One, Blameless One, Constant Love.*

24 <answersingenesis.org/go/theyhadnames/>
25 ibid.
26 McCutchen 1993, p. 90.

Intriguingly, the *Red Record* seems to trace its own origin to this time:

> The next sachem was History Man; written records he began.[27]

Also during this era, the Algics appear to have adopted a practice conducive to extensive population growth—farming:

> The next sachem was Hominy Man; raising crops he began.[28]

Immediately following *Hominy Man* was a sachem called *Subdivider.*[29] At this stage, the second Algic population split occurred, giving rise to the Blackfeet (see **Appendix B**). The *Red Record* description suggests this was an amiable split.

The Blackfeet's own history, and especially the geography of this history, supports this narrative. During the time of *Subdivider*, the main body of Algics was located near Idaho/western Montana, as per McCutchen (**Figure 26**). It's also not far from what the Blackfeet called their ancestral homes:

> During the seven years I [D.D. Mitchell] spent amongst [the Blackfeet], I frequently amused myself by collecting such historical information as the old *savans* of the tribe had learned from their forefathers; together with their own recollections of more modern times. From all I could learn, the Blackfeet originally inhabited that region of country watered by the Sascatchawain [Saskatchewan] and its tributaries, never extending their hunting or war parties farther south than the head waters of the Marias river, or farther east than the head waters of the Milk river…[30]

The Saskatchewan River and its branches flow from west to east through modern Alberta, Saskatchewan, and Manitoba (**Figure 36**). The Marias River flows through northern Montana into the Missouri River (**Figure 36**).

27 McCutchen 1993, p. 90.
28 McCutchen 1993, p. 94.
29 McCutchen 1993, p. 94.
30 Schoolcraft 1855, p. 685. I retained original spellings and conventions.

> The cause of their separation and dispersion over a wider range of country grew out of a civil war regarding the claims of two ambitious chiefs, each claiming sovereign powers. ... The warriors being divided, enrolled themselves under the two banners; the younger and more warlike, under the red; the old men contending for the hereditary claims of the black chief. After many skirmishes and assassinations, a pitched battle ensued, which resulted in the disastrous defeat of the black chief. ... After the defeat of the black party, they fled towards the south, still marching under their black banner, and clothed in deep mourning. They appear to have reached the Missouri during the fall, when the prairies were burning; and the black ashes of the burnt grass had coloured their moccasins and leggings. In this condition, they were first seen by the Crow Indians, who called them Blackfeet.[31]

Thus, the ancestral Blackfeet took up residence just east of where the Algics were when the population split occurred.

◂◂ ▪ ▸▸

After about a century of peace, conflict resumed for several decades (i.e., approximately A.D. 1095 to A.D. 1155).[32] Once again, it was due to outside forces—just not the human kind. *Drought* was both the environmental stressor for the nation as well as the name of the presiding sachem. The lack of water pushed the Algic eastward onto the North American Great Plains.

Internal conflict divided the nation again:

> After [the Hardened One] was the Denouncer, rebellious and unwilling.
>
> In anger, leaving eastward, some went away in secret.[33]

Linguistically, this split would correspond to the Cree/Arapaho/Gros Ventre language subgroup (see **Appendix B**).

---

31 Schoolcraft 1855, p. 685-686. I retained original spellings and conventions.

32 McCutchen 1993, p. 94.

33 McCutchen 1993, p. 94.

The geography of this split underscores this identification. At this point, the main body of Algics was located in the borderlands of what is now Montana and Wyoming, as per McCutchen (**Figure 26**). Compare this location to the stated homelands of the Cree:

> The Plains Cree, along with their Assiniboine allies, were the last tribe to move onto the Plains in historic times. … Cree speakers moved west as a result of game depletion in the area east of Lake Winnipeg, resulting from the fur trade in the late eighteenth and early nineteenth centuries.[34]

Lake Winnipeg sits in the middle of modern Manitoba, to the northeast of Montana/Wyoming (**Figure 36**). Thus, the Cree may have once resided to the east of Manitoba in modern Ontario.

Now compare the location of the main body of Algics to the stated homelands of the Arapaho/Gros Ventre:

> The Arapaho, with their subtribe, the Gros Ventres, are one of the westernmost tribes of the wide-extending Algonquian stock. According to their oldest traditions they formerly lived in northeastern Minnesota and moved westward in company with the Cheyenne, who at that time lived on the Cheyenne fork of Red river. From the earliest period the two tribes have always been closely confederated, so that they have no recollection of a time when they were not allies. In the westward migration the Cheyenne took a more southerly direction toward the country of the Black hills, while the Arapaho continued more nearly westward up the Missouri. The Arapaho proper probably ascended on the southern side of the river, while the Gros Ventres went up the northern bank and finally drifted off toward the Blackfeet, with whom they have ever since been closely associated, although they have on several occasions made long visits, extending sometimes over several years, to their southern relatives, by whom they are still regarded as a part of the "Inûna-ina." The others continued on to the great divide between the waters of the Missouri and those of the Columbia, then turning southward along the mountains,

34 Darnell, R., "Plains Cree," p. 638, 640, in: Sturtevant 2001.

> separated finally into two main divisions, the northern Arapaho continuing to occupy the head streams of the Missouri and the Yellowstone, in Montana and Wyoming, while the southern Arapaho made their camps on the head of the Platte, the Arkansas, and the Canadian, in Colorado and the adjacent states, frequently joining the Comanche and Kiowa in their raids far down into Mexico[35] [see **Figure 37** for a map of these migrations].

It's not hard to imagine the original group of Cree/Arapaho/Gros Ventre ancestors leaving the rest of the Algics in Wyoming/Montana, migrating east through the Dakotas to Minnesota, and then the Cree branching off and going north.

◂◂ ▪ ▸▸

The national separation quelled the anger among the remaining body of Algics. The *Red Record* mentions subsequent sachems like *Beloved One* and *Tamanend*, who was "everyone's friend."[36] Almost 80 years passed during this period (i.e., roughly A.D. 1130 to A.D. 1215).

Then, in the early A.D. 1200s, troubles from the outside re-entered the scene:

> White Chick was the sachem; once more, bloodshed, north and south.
>
> Wise and crafty in council, Mighty Wolf was the sachem.
>
> He could fight every foe; the Strong Stone he struck down.
>
> Whole-Hearted was the sachem fighting the Snakes.
>
> Strong is Good was the sachem fighting the North Walkers.
>
> Poor One was the sachem fighting invaders.[37]

One of these enemies has a very descriptive name, the *North Walkers.* Who were they? Given the Algic origin in the north, perhaps it was a group they had encountered earlier in their journey.

---

35 Mooney 1896, p. 954. I retained original spellings and conventions.
36 McCutchen 1993, p. 98.
37 McCutchen 1993, p. 102.

In a previous section, we observed that a group of Athabaskans did indeed migrate from the north and end up along the Mexican border at Contact. Since this period was pre-European, before the time when Europeans brought their horses, I think we can safely assume that the northerly Athabaskans went south by *walking*.

The Mescalero Apache put a time stamp on these migrations:

> the beginning of this southward migration was also the time of their separation from the Slave Indians [see **Figure 31** for Contact location of Slavey], estimated at about 600 years ago.[38]

This information was obtained from a Mescalero Apache elder in A.D. 1985. Thus, the Mescaleros began their migrations six centuries earlier, in the late A.D. 1300s.

Linguistically, the Mescaleros were but one of several Athabaskan groups in the southwest at Contact (**Figure 29**). The southwestern groups form their own linguistic subcluster, though each group within the subcluster is distinct. Apparently, some of the latter migrated separately:

> Mescaleros never made pottery, and they were the last of the Apache groups to migrate south. They were preceded in the Southwest by three earlier groups: hunting peoples, those who made pottery, and the small 'ancient people' (*ńda saane*).[39]

When did these three earlier migrations occur? The last (fourth) migration was that of the Mescalero. Their journey began in the late A.D. 1300s (see previous paragraphs). It follows, then, that the first three migrations happened earlier—perhaps in the early A.D. 1300s, if not the A.D. 1200s.

When the Algic fought the *North Walkers*, the estimated date range is A.D. 1214 to A.D. 1243. This time frame could plausibly be connected to the time frame of the first of the Athabaskan migrations south. This suggests that the Algics did battle with the Eyak-Athabaskans not

38 Carmichael, David L. and Farrer, Claire R., "We do not forget; we remember: Mescalero Apache origins and migration as reflected in place names," p. 187, in: Seymour 2012.

39 Ibid., p. 186.

once but twice. The first time was in far northwest North America; the second, on the Great Plains (**Figure 38**).

◂◂▪▸▸

The Athabaskans weren't the only enemies with which the Algic had to contend. Other sachems had to fight off the "Strong Stone," "Snakes," and "invaders" (see stanzas in previous section). These descriptors aren't as evocative as "North Walkers." But the general geography of the battle sites—the northern Great Plains—suggests some candidates for the identity of these foes.

At Contact, the *Kiowa*, a branch of the Kiowa-Tanoan language family, were on the southern Great Plains (**Figure 39**). But their own history[40] puts their origins farther north:

> Unlike the neighboring Cheyenne and Arapaho, who yet remember that they once lived east of the Missouri and cultivated corn, the Kiowa have no tradition of ever having been an agricultural people or anything but a tribe of hunters. The earliest historic tradition of the Kiowa locates them in or beyond the mountains at the extreme sources of the Yellowstone and the Missouri, in what is now western Montana.[41]

This is the region that the Algic had just passed through before encountering the "Strong Stone," "Snakes," and "invaders" (**Figure 26**).

> They describe it as a region of great cold and deep snows, and say that they had the Flatheads (*A'dalton-kd-igiha'go*, "compressed head people") near them, and that on the other side of the mountains was a large stream flowing westward, evidently an upper branch of the Columbia.[42]

The *Flatheads* were part of the Salish language family (**Figure 32**) and, at Contact, were just south of the Columbia River.

> Here, they say, while on a hunting expedition on one occasion, a dispute occurred between two rival chiefs over the possession of the udder of a female antelope, a delicacy

40 Mooney 1898, p. 153-154. I retained original spellings and conventions.
41 Mooney 1898, p. 153.
42 Mooney 1898, p. 153.

> particularly prized by Indians. The dispute grew into an angry quarrel, with the result that the chief who failed to secure the coveted portion left the party and withdrew with his band toward the northwest, while the rest of the tribe moved to the southeast, crossed the Yellowstone (*Tsosd P'a*, "pipe (?) stone river"), and continued onward until they met the Crows (*Gad-k'iago*, "crow people"), with whom they had hitherto been unacquainted.[43]

At Contact, the Crow territory included the upper part of the Yellowstone River (**Figure 39**).

> For a while they continued to visit the mountains, but finally drifted out into the plains, where they first procured horses and became acquainted with the Arapaho and Cheyenne, and later with the Dakota. Keim, writing in 1870, says that the Kiowa "claim that their primitive country was in the far north," from which they were driven out by wars, moving by the aid of dogs and dog sledges. "From the north they reached a river, now the south fork of the Platte [see **Figure 39** for geography of the rivers in this paragraph]. Their residence upon this river is within the recollection of the old men of the tribe. Not satisfied with the Platte country, they moved on across the Republican and Smoky Hill rivers until they reached the Arkansas. Thence they moved upon the headwaters of the Cimarron. Here they permanently located their council fire, and after much lighting secured control of all the country south of Arkansas river and north of the Wichita mountains and headwaters of Red river."
>
> There can be no doubt as to the correctness of the main points of this tradition, which is corroborated by the testimony of the northern Arapaho and other tribes of that region.[44]

This latter part of their journey (**Figure 39**) takes them away from the Algic battlegrounds of the early A.D. 1200s. But the earlier part was in the right location. The only question is the timing of when they were located there.

43 Mooney 1898, p. 153.
44 Mooney 1898, p. 153-154.

Whatever the exact identity of these foes, the outcome was unambiguous. Decades of bloodshed produced in the minds of the Algics a strong desire to escape *by going east* (estimated date range of A.D. 1229 to A.D. 1258):

> East Looking was the sachem, melancholy about the war there was.
>
> "To the Rising Sun now you must go," he said; many were those who eastward went.
>
> They separated at the Mississippi; the lazy ones remained behind.[45]

By the linguistic sequence established in **chapter 4** (see also **Appendix B**), this division of Algics gave rise to the Menominee and Cheyenne.

At Contact, the Menominee were in Wisconsin (**Figure 25**), not far from the headwaters of the Mississippi in next-door Minnesota. The Cheyenne were on the Plains (**Figure 25**). But that's not where they started. The Arapaho traditions put the Cheyenne origins farther east (see quote in previous section). The Cheyenne themselves said the same thing:

> From their own traditions and from the fact that they are of the western Algonquians, whose current of migration was in a general westerly and southwesterly course, we may conclude that the Cheyenne came from the Northeast or East, but whether they reached the upper drainage of the Mississippi River along the southern shores of the Great Lakes, or along the northern shores and around the west end of Lake Superior [see **Figure 37**], it would seem now impossible to learn.[46]

The Cheyenne and Menominee made out better than the main body of Algics. At the time of separation, the latter were approaching the Mississippi River. In short order, they ran into an even bigger obstacle than the ones they faced on the Plains—with tragic results.

---

45 McCutchen 1993, p. 102, 106.

46 Grinnell 1923, p. 4. I retained original spellings and conventions.

# 6

## The Empire
## Late A.D. 1200s

Late in the A.D. 1200s, in the temperate climate just below the confluence of the Missouri and Mississippi Rivers where modern St. Louis now stands, the Algics sought relief from the conflicts that had harassed them on the wide-open Great Plains. The American Bottom would have been a prime location to settle. Soil was fertile. Game was abundant. But someone else had gotten there first. Someone powerful.

Up to this point in the *Red Record*, the Algics had employed a variety of descriptive names for their American enemies: *Snakes, Charlatans, Blackened Ones, Stony Ones, Strong Stone, North Walkers, Invaders.*[1] They had narrated times of *much evil, bloodshed*, and *war.*[2] But at the Mississippi, they described something categorically different (est. date range of A.D. 1236 to A.D. 1266):

> The Lodge Man was the sachem; the Talega possessed the East.[3]

*Possessed the East?* Even if we limit our focus to modern U.S. lands, *east* of the Mississippi River covers an area of more than half a million square miles. Apparently, the *Talegas* were the owners of it. They were *empire builders.*

1 McCutchen 1993, p. 82, 86, 102.
2 McCutchen 1993, p. 90, 102.
3 McCutchen 1993, p. 106.

The Algics didn't even attempt a fight. Instead, they opted for a more diplomatic approach (est. date range of A.D. 1243 to A.D. 1273):

> Strong Ally was the sachem; land to the east he asked for.[4]

Unfortunately, the Possessors of the East were none too eager to surrender their lands:

> Eastward some traveled; the Talega king massacred them.[5]

◂◂ ▪ ▸▸

Who were these murderous new enemies? The ruins near and east of the Mississippi River hint at an answer—an answer that reveals one of the most significant chapters in all of Native North American history. If there was ever an analogy to the Roman Empire in North America, this was the period. It will be worth our time to work through the details slowly and carefully.

Today, just east of St. Louis, sits the biggest earthwork mound anywhere in the Americas. *Monk's Mound* (**Figure 40**) doesn't look like a product of the Hopewell culture. Yes, it's geometric, but it doesn't consist of a set of earthen walls surrounding an empty center. The center itself forms most of the mound.

It also doesn't bear any resemblance to the products of the Effigy culture. The shape of the mound itself is more a square or rectangle than anything resembling an animal or person.

Perhaps the biggest signals of change are the height of the summit at 100 feet[6] and the shape of the mound: flat-topped.

It wasn't just St. Louis that sported the distinctive Monk's Mound style. Eventually, pyramid-like, flat-topped mounds would appear all the way from Wisconsin (**Figure 41**) to Alabama (**Figure 42**) to Georgia (**Figure 43**).

The last time we saw a change in mound-building culture (from Hopewell to Effigy), it marked the arrival of an invader (see **chapter 3**). The shift in mound-building culture from Effigy to *Mississippian* was no different.

---

4 McCutchen 1993, p. 106.
5 McCutchen 1993, p. 106.
6 According to signage on site.

In 1836, when the earliest European explorers first saw the biggest flat-topped mound site in Wisconsin, they couldn't help but let their imaginations wander elsewhere:

> Along the Crawfish River, fifty miles west of the new village of Milwaukee, the newcomers stumbled upon the burned ruins of a massive fortification enclosing flat-topped earthen mounds. ... The settlers were reluctant to believe that native North Americans built these walls and mounds. Instead, the similarity of the site to the better known Aztec ruins in Mexico led to the conclusion that the Aztecs were involved and the site must be that of Aztalan or Aztlán, the Aztec northern homeland. The name stuck.[7]

And why not? Pre-Columbian Mesoamerica witnessed flat-topped mound construction for over a millennium. In central Mexico, the residents of Teotihuacan erected massive flat-topped pyramids and temples of stone in the first few centuries after Christ (**Figure 22**). In Guatemala and southern Mexico, the Classic Maya built one flat-topped temple after another during most of the first millennium A.D. and later. (**Figure 44**). Closer to Contact, the Aztecs made flat-topped temples. Reconstructions reveal their grandeur (**Figure 45**).

These visual hints turned out to be more than a coincidence.

◂◂ ▪ ▸▸

In 1774, a Frenchman by the name of Le Page Du Pratz recorded[8] a Native American nation's history—a history that resonated with much of what we've just observed. When Du Pratz spoke to them, they were north of the Rio Grande, in what is now Louisiana. But according to one of the tribe's temple keepers, their original homes were farther south:

> "Before we came into this land we lived yonder under the sun, (pointing with his finger nearly south-west, by which I understood that he meant Mexico;) we lived in a fine country where the earth is always pleasant; there our Suns [leaders] had their abode, and our nation maintained itself for a long time against the ancients of the country, who

7 Birmingham and Goldstein 2005, p. ix-x.
8 Du Pratz 1774.

> conquered some of our villages in the plains, but never could force us from the mountains."[9]

Even while in Mexico, this nation ruled a large territory, as per the temple keeper:

> "Our nation extended itself along the great water where this large river loses itself."[10]

What was *this large river*? Du Pratz was speaking to a man in what is now Louisiana. The largest river in Louisiana is the Mississippi River. The Mississippi loses itself—empties itself—into the Gulf of Mexico. This implies that the ancestors of this tribe lived all the way from the mountains of Mexico along the western shores of the Gulf of Mexico[11] to modern Louisiana (**Figure 46**).

This originally Mesoamerican tribe expanded even farther:

> "But as our enemies were become very numerous, and very wicked, our Suns sent some of their subjects who lived near this river, to examine whether we could retire into the country through which it flowed."[12]

The homelands of this tribe went along the Gulf of Mexico all the way up to the Mississippi delta. It would seem, then, that the request of the Suns (leaders) was to explore the land of the Midwest surrounding the Mississippi.

> "The country on the east side of the river being found extremely pleasant, the Great Sun, upon the return of those who had examined it, ordered all his subjects who lived in the plains, and who still defended themselves against the antients [ancients?] of this country, to remove into this land, here to build a temple, and to preserve the eternal fire. A great part of our nation accordingly settled here, where they lived in peace and abundance for several generations."[13]

---

9 Du Pratz 1774, p. 279-280.

10 Du Pratz 1774, p. 279-280.

11 With respect to the Gulf, they were the western shores; with respect to Mexico, they were the eastern shores.

12 Du Pratz 1774, p. 280.

13 Du Pratz 1774, p. 280.

At this point in their history, the majority of the nation appears to have moved to east of the Mississippi—the area of the southeastern United States and perhaps of the eastern woodlands (**Figure 47**).

> "The Great Sun, and those who had remained with him, never thought of joining us, being tempted to continue where they were by the pleasantness of the country, which was very warm, and by the weakness of their enemies, who had fallen into civil dissentions, in consequence of the ambition of one of their chiefs, who wanted to raise himself from a state of equality with the other chiefs of the villages, and to treat all the people of his nation as slaves. During those discords among our enemies, some of them even entered into an alliance with the Great Sun, who still remained in our old country, that he might conveniently assist our other brethren who had settled on the banks of the Great Water to the east of the large river, and extended themselves so far on the coast and among the isles, that the Great Sun did not hear of them sometimes for five or six years together."[14]

In this paragraph, the context for the phrase "among the isles" is the geographic reach of the members of the nation who had left Mexico. The coast of the Gulf of Mexico goes eastward from the Mississippi delta along the coast of Florida (**Figure 47**). Did the tribe make it to the isles of the Florida Keys? To the islands of the Caribbean? Either of these possibilities would be extraordinary. The distance would also be consistent with the statement that the Great Sun didn't hear from them "sometimes for five or six years together."

> "It was not till after many generations that the Great Suns came and joined us in this country, where, from the fine climate, and the peace we had enjoyed, we had multiplied like the leaves of the trees."[15]

By this point, it seems that this Native American nation had achieved a large population size.

14 Du Pratz 1774, p. 280.
15 Du Pratz 1774, p. 280.

When was this? When did the Great Suns join the rest of their subjects? The description immediately following supplies the answer:

> "Warriors of fire, who made the earth to tremble, had arrived in our old country, and having entered into an alliance with our brethren, conquered our ancient enemies; but attempting afterwards to make slaves of our Suns, they, rather than submit to them, left our brethren who refused to follow them, and came hither attended only with their slaves."
>
> Upon my asking him who those warriors of fire were, he replied, that they were bearded white men, somewhat of a brownish colour, who carried arms that darted out fire with a great noise, and killed at a great distance; that they had likewise heavy arms which killed a great many men at once, and like thunder made the earth tremble; and that they came from the sun-rising in floating villages.
>
> The ancients of the country he said were very numerous, and inhabited from the western coast of the great water to the northern countries on his side the sun, and very far upon the same coast beyond the sun. They had a great number of large and small villages, which were all built of stone, and in which there were houses large enough to lodge a whole village. Their temples were built with great labour and art, and they made beautiful works of all kinds of materials.[16]

Clearly, the "warriors of fire" were the Spanish conquistadors. At Contact, the peoples of Mesoamerica were indeed very numerous, and the Aztecs were well-known for their temples, great number of large and small villages, and beautiful works. Therefore, the Great Suns didn't leave Mexico until the A.D. 1500s.

Since this was the second of two migrations, we can therefore conclude that the first migration from Mexico happened at some undetermined time prior to the A.D. 1500s.

Before Contact, this nation must have been a sight to behold:

---

16 Du Pratz 1774, p. 280-281.

> [A]ccording to their traditions they were the most powerful nation of all North America, and were looked upon by the other nations as their superiors, and on that account respected by them. To give an idea of their power, I shall only mention, that formerly they extended from the river Manchac, or Iberville, which is about fifty leagues from the sea, to the river Wabash, which is distant from the sea about four hundred and sixty leagues; and that they had about five hundred Suns or princes.[17]

The Iberville River today is found in Louisiana; the Wabash, in modern Indiana (**Figure 48**). To rule from Louisiana to Indiana is to rule a massive tract of land.[18] This is consistent with the additional claim that they "had about five hundred Suns or princes." A kingdom that widespread would seem to require a large body of supervisors to maintain.

In 1823, John Haywood suggested that the domains may have encompassed even more land, from

> the river Manches, or Iberville, which is about 50 leagues from the sea, to the Wabash, which is about 450 leagues from the sea; **and it is probable, that they extended laterally up all the rivers which fall into Mississippi between these two extremes**. The mounds are perhaps within the limits of their settlements, and not beyond them. They had at this time 500 sachems of the nation[19] [emphasis added].

Technically, the Missouri River in the American west falls into the Mississippi. But the focus of both Haywood and Du Pratz seems to have been east of the Mississippi. This is where the Wabash flowed,

17 Du Pratz 1774, p. 299.

18 In the Du Pratz quote, he describes distances from the Iberville to the Wabash. By modern measurements, the Iberville is about 40 miles as the crow flies from Lake Pontchartrain in Louisiana, and only around 80 miles from the Gulf coast of Louisiana. The mouth of the Wabash is over 500 miles straight line distance to the Gulf Coast. Du Pratz doesn't report the distance in miles but in leagues—the Iberville being "about fifty leagues from the sea" and the Wabash being "distant from the sea about four hundred and sixty leagues." By standard encyclopedia definitions, a league represents "2.4 to 4.6 statute miles" (britannica.com/science/league-measurement). Therefore, "fifty leagues from the sea" would seem to be 120 to 230 miles, not 80 miles. The distance of "four hundred and sixty leagues" would seem to be 1100 to 2100 miles, not 500 miles. I don't know why Du Pratz reported such inflated distance values.

19 Haywood 1823, p. 99. I retained original spellings and conventions.

eventually connecting with the Ohio River, which empties into the Mississippi (**Figure 48**). The eastern Mississippi watershed is vast (**Figure 49**).

It seems, then, that the Natchez, the nation explicitly identified in Du Pratz's writings (see also **Figure 50** for their location in Louisiana at Contact), once ruled much of eastern North America (**Figure 51**). They *possessed the east.*

◂◂▪▸▸

When did the Natchez first begin to possess the east? So far, we've been able to derive part of the answer—sometime before the A.D. 1500s. But for the Natchez to be viable candidates for the Talegas, this isn't good enough. They need to have been there at least by the A.D. 1200s.

Were they? Archaeological correlations, specifically to the archaeologically defined Mississippian culture, supply the answer.

Like the Natchez, Mississippian chiefdoms eventually spread across much of the southeast United States (**Figure 52**). Do you see the large amount of overlap between what the Natchez claimed as their territory (**Figure 51**), and where the Mississippian cultures were once found (**Figure 52**)? Did the spread of *chiefdoms* also remind you of the Natchez claim to once having had *500 Suns? Mississippian* and *Natchez* seem nearly synonymous.

When did the Mississippian culture arise? The full archaeological context reveals the dates. Before the Mississippian period, the preceding *Late Woodland* archaeological period was unremarkable, a "dark age." There were "few large villages or ceremonial centers and every indication of highly localized, if territorial, lifeways."[20]

But then a transformation occurred:

> After ca. A.D. 900, societies characterized by central communities with temple/mortuary mounds arranged around plazas, hereditary inequality between people and groups, and a reliance on intensive maize agriculture began to

20 Alt, S.M., "Making Mississippian at Cahokia," p. 499, in: Pauketat 2012.

> appear in the Southeast. … Over the next several centuries, similar societies appeared across much of [the] region.[21]

In other words, Mississippian (Natchez) culture arose around the A.D. 900s. By the A.D. 1200s, Mississippian culture was spread all around the *east* (**Figure 52**)—just in time for the Algic approach from the west.

Thus, the Mississippian (Natchez) fulfill the description of the Talegas.

◂◂ ▪ ▸▸

The Natchez may not have been the sole builders of the Mississippian empire. They certainly were not the only Native Americans north of the Rio Grande who claimed origins in Mexico. At Contact, the warm lands of the southeast, the lands which were once the domain of the Natchez, were claimed by a group of nations related under the *Muskogee* language family. Among them were the Alabama, Apalachee, Choctaw, Chickasaw, and Creek (**Figure 53**).

In 1802, Louis LeClerc de Milford[22] published an account of what the Creeks related about their origins:

> After the Spanish conquest of Mexico, all the world knew that this beautiful land of North America was inhabited by a docile and peaceable people who, ignorant of firearms, were easily subjugated. They had only courage and numbers to oppose to the deadly weapons of their enemies; in short, they were defenseless. For what availed bows and arrows against the artillery of an army, weak indeed in numbers, but inured to war, intrepid, and led on by an insatiable greed of gold, which this too trusting people had been unfortunate enough to parade before their eyes? Montezuma then reigned in Mexico; finding that he was unable to check the progress of the Spanish, he summoned to his aid the tribes contiguous to his dominions. The Muskogee nation, now known as Creeks, which formed a separate republic in the northwestern part of Mexico and had a formidable number of warriors, offered him assistance, an assistance that would

21 Anderson and Sassaman 2012, p. 155.
22 de Milford 1956. I retained original spellings and conventions.

> have been redoubtable for any but a disciplined army such as that of the Spanish under Hernando Cortes. The courage of this martial people only served to effect its speedier destruction and could not save Montezuma, who lost his life, and his empire, which was almost totally depopulated. After the death of Montezuma and several other chiefs, the Muskogees, considerably weakened by this dreadful war which they were no longer in a position to wage, chose to abandon a country that offered them in exchange for their past happiness only the most terrible slavery and to seek another that would ensure them the ample resources and the peace and tranquillity of which the Spanish had just despoiled them.[23]

Though the Creeks also traced their origins to Mexico, their account contrasts to that of the Natchez on several points. Unlike the Natchez, the Creeks aided the Aztecs against the Spanish before fleeing; the Natchez aided the Spanish against the Aztecs before fleeing. The Creeks also identified the "northwestern part of Mexico," not the Gulf coast of Mexico, as their homelands. Finally, the Creek account makes no mention of an earlier, pre-Conquistador migration.

This would suggest that the Muskogeans were *not* co-builders of the Mississippian empire.

But the Creek account contains a key detail suggesting a more nuanced interpretation. After describing their migration north to the Red River, they moved east across the prairies of what is now Texas, eventually reaching a forest, where they settled down and planted corn.[24] Then they were forced to move again—by an enemy whose identity raises several questions:

> They were discovered by the Alibamus, who killed several of their people. So the elders, the natural chiefs of the nation, called together the young warriors and sent them on the trail of the assassins, but without success, because there

23 de Milford 1956, p. 161-162.
24 de Milford 1956, p. 162-165.

> was no coordination in their operations and they lacked a chieftain.[25]

I presume that the *Alibamus* were the same people that we call the *Alabamas* today—the linguistic relatives of the Creeks. Did the Alabamas also originate in Mexico but leave before the Creeks did? The passage above suggests that, at a minimum, the Creeks and Alabamas were not escaping as a single, unified group.

At some point, the Creeks must have crossed the Mississippi. At Contact, they were found in what is now Georgia and Alabama (**Figure 53**).

What about the other linguistic relatives of the Creeks? Do they contain any clues as to from where and when they originated, especially with respect to the Natchez? The Choctaw (or *Cha'hta*) also describe a west-to-east migration across the Mississippi:

> The Cha'hta trace their *mythic origin* from the 'Stooping, Leaning or Winding Hill,' Náni Wáya, a mound of fifty feet altitude, situated in Winston county, Mississippi, on the headwaters of Pearl river. The top of this 'birthplace' of the nation is level, and has a surface of about one-fourth of an acre. One legend states, that the Cha'hta arrived there, after crossing the Mississippi and separating from the Chicasa, who went north during an epidemic.[26]

This passage implies that the Choctaw and Chickasaw (*Chicasa*) migrated as one body until the epidemic. Perhaps the Creeks were also part of this group. Another early 1800s account says so explicitly:

> The account of the *Cha'hta* migration, as given in the Missionary Herald, of Boston, Vol. XXIV (1828), p. 215…"When they emigrated from a distant country in the west, the Creeks were in front, the Cha'hta in the rear. They travelled to a 'good country' in the east; this was the inducement to go. On the way, they stopped to plant corn. Their great leader and prophet directed all their movements, carried the *hobuna* or sacred bag (containing "medicines") and a long white pole as the badge of his

25 de Milford 1956, p. 165.

26 Gatschet 1884, p. 105. I retained original spellings and conventions.

> authority. When he planted the white pole, it was a signal for their encampment. He was always careful to set this pole perpendicularly and to suspend upon it the sacred bag. None were allowed to come near it and no one but himself might touch it. When the pole inclined towards the east, this was the signal for them to proceed on their journey; it steadily inclined east until they reached Nánni Wáya. There they settled."[27]

The Chickasaw have their own, familiar-sounding, migration account:

> By tradition, they say they came from the West; a part of their tribe remained in the West. When about to start eastward, they were provided with a large dog as a guard, and a pole as guide; the dog would give them notice whenever an enemy was near at hand, and thus enable them to make their arrangements to receive them. The pole they would plant in the ground every night, and the next morning they would look at it, and go in the direction it leaned. They continued their journey in this way until they crossed the great Mississippi River; and, on the waters of the Alabama River, arrived in the country about where Huntsville, Alabama, now is: there the pole was unsettled for several days; but, finally, it settled, and pointed in a southwest direction. They then started on that course, planting the pole every night, until they got to what is called the Chickasaw Old Fields, where the pole stood perfectly erect. All then came to the conclusion that that was the Promised Land. ... When they were travelling from the West to the Promised Land in the East, they had enemies on all sides, and had to fight their way through, but they cannot give the names of the people they fought with while travelling. They were informed, when they left the West, that they might look for whites; that they would come from the East; and they were to be on their guard, and to avoid the whites, lest they should bring all manner of vice among them.[28]

27 Gatschet 1884, p. 219-220.
28 Schoolcraft 1851, p. 309-310. I retained original spellings and conventions.

What *whites* might the Chickasaw be looking for? What *whites* were coming "from the East"? Surely this refers to Europeans. If so, then the timing of the migration from "the West" must be A.D. 1500s or later—which is exactly in line with the Creek account.

The similarities among the Chickasaw, Choctaw, and Creek accounts suggest that all three were describing the same thing. It also, again, suggests that the Creek, Choctaw, and Chickasaw were *not* participants in the origin and growth of the Mississippian empire. But are their accounts plausible? Shortly after Cortés landed in Mexico in 1519, presumably prompting the migrations of the Muskogeans, other Europeans began to explore what is now the United States. In 1528, Pánfilo de Narváez encountered the *Apalachee*, speakers of a Muskogean language, along the Gulf coast of what is now Florida (**Figure 53**).[29] From 1539 to 1543, Hernando de Soto marched his troops through much of the southeast of what is now the United States (**Figure 54**). His accounts included a mention of a people termed *Chicaza*—presumably, the *Chickasaw*.

I suppose that it's possible that, in just a few years' time, the Creek/Choctaw/Chickasaw had left Mexico, arrived in the southeast, and had set up villages—villages large enough that European explorers marked their names in their journals. In fact, their own accounts suggest a migrating population of "ten thousand men, besides women and children."[30] And yet…the timing of this sequence of events seems so compact.

Perhaps there was more going on than these accounts have explicitly described. Perhaps there are additional Muskogean accounts that we have not yet recovered. For one, we have the hint that the Alabamas may have preceded the main body of Choctaw/Chickasaw/Creeks. For another, we have two major groups of Native Americans describing a migration from Mexico—the Natchez and Muskogeans. The former detail at least two rounds of migrations. Perhaps the Muskogeans had two rounds as well—one at Contact and one prior to it, perhaps centuries prior to it—yet not recorded in the accounts I cited above.

29 Hudson and Tesser 1994, p. 10, 157.
30 Williams 2022, p. 205 [page numbers as per the pagination on right side of webpage].

Or...perhaps the Natchez and Muskogeans ultimately derived from the same ancient source population in Mexico, separating later in pre-Columbian Mexican history:

> Broader connections of Muskogean with other language groups of the Southeast have been proposed. ... The possible connection between Natchez and Muskogean deserves investigation.[31]

In the distant past, I suspect that the Muskogeans and Natchez were a single people group. If so, then the initial Natchez migration may have contained both Natchez and Muskogean relatives (e.g., the Alabamas). Then, at some later point, those Muskogeans and Natchez who had remained in Mexico went their separate ways, leading to the differing post-Conquistador migration accounts.

◂◂ ▪ ▸▸

In mid-March of 2024, I took a trip to *Cahokia*, the site containing Monk's Mound, and gained a new perspective on the tragic meeting between the Algics and the Talegas (Natchez, and perhaps Muskogeans) 800 years prior. I learned that *tragic* is but the beginning of the adjectives appropriate to this fatal encounter.

I entered the grounds for the Cahokia Mounds State Historic site from the east. Immediately in front of me was *Monk's Mound*. Wow. The biggest earthwork in all of the Americas is even more impressive in person.

But Monk's Mound wasn't my primary focus that day. Prior to the trip, I had known that Cahokia was a destination in its own right, independent of Monk's Mound. The largest pre-Columbian city north of the Rio Grande, Cahokia had once boasted a population of 10,000 to 15,000 people.[32] In its day, it was as large as London.[33] I had planned to walk the entire grounds to experience the context, the atmosphere of this unique set of ruins.

I quickly turned left down the entrance road to the Interpretive Center parking lot. I found the starting point for the 10-mile trail and

---

31 Campbell 1997, p. 149.
32 Alt, S.M., "Making Mississippian at Cahokia," p. 500, in: Pauketat 2012.
33 https://www.britannica.com/place/London/History

set out. It was along this path that clues to the nature of Cahokian culture began to emerge.

The trail took me through all four corners of the park, even the most forsaken areas. It was at these distant spots that the *city* element of Cahokia hit me. There's no way, I thought, someone on the top of Monk's Mound could yell and expect people to respond in the southwest and southeast corners of the site.[34] There's no way to communicate with and control the different sections of the city apart from some centralized administrative structure. You need a system—with people responsively running it—to link the parts of the city and keep them humming together.

In other words, to run a centralized administration, you need some measure of *power*.

I got another sense for *city* the next day in the car. Collinsville Road splits the Cahokia site into two parts. I approached Cahokia on Collinsville Road from the west, and it didn't take long before the breadth of the location became clear. From west to east, the park runs more than a mile. Even driving at 45 mph, I was struck by how the grassy fields just kept going. In its prime, Cahokia's footprint was more than 4.5 square miles, or around 3,000 acres. The city also contained over 120 mounds.[35]

Back to the trail on my first day there, the wandering 10-mile path eventually took me back to Monk's Mound. My initial approach was from the east (**Figure 55**). In hindsight, I realized that this was one of the best directions from which to come. The mound itself and the steps to ascend it face south. But the southern side of Monk's Mound is broken up into several levels, which makes the overall slope seem less steep when you're at the bottom. Similarly, the western side also rises in stages. The northern side has trees, perhaps to prevent erosion and

---

34 I've treated these corners of the park as part of the city proper. Technically, this is still a matter of debate among professional archaeologists. This was indirectly confirmed to me on my second day at Cahokia. While I was standing on the top of Monk's Mound, I saw someone pulling a piece of equipment in the northeast corner of the site; I thought it might be a remote sensing device. Being a nerd, I hopped down the mound to investigate. The PhD running the equipment graciously gave me some of his time, confirmed that it was indeed remote sensing equipment, and that he was a participant in an on-going multi-university research project to map out Cahokia's full urban scale.

35 Alt, S.M., "Making Mississippian at Cahokia," p. 497, in: Pauketat 2012.

collapse. The trees partially obstruct the view. On the eastern side, the angles of the mound are uninterrupted, steep, and free of foliage.

Standing at the base and gazing 100 feet in the air, I imagined what it may have been like 800 years ago as a commoner who was summoned to see the chief or leader at the top. The psychological effect reminded me of an experience I had in Germany. My family and I were on a trip to visit King Ludwig's *Neuschwanstein* castle in Bavaria, the same castle on which the iconic Disney World® castle is based. On the tour of the rooms inside Neuschwanstein, we were led into the throne room. The passageway was a dim but open-sided hall with a low ceiling. The end of the hall opened to a bright, elevated throne on my right backed by a high, glittering wall and domed ceiling. Gold was everywhere. I immediately felt what was surely the desired emotional impact—I was small and low and insignificant; the king was high and majestic and exalted above me.

Back at Cahokia and looking up the mound, I felt a similar sensation greet me. Had I been summoned back then, the ruler of Cahokia would have been so high; I would have been so low and irrelevant. The unambiguous message from the chief to someone like me was, *I'm much more powerful than you. I'm in charge; you're not*. Perhaps *king* rather than *chief* would have been a better descriptor of the one who ruled from the summit of Monk's Mound.

As I ascended the steps on the south, I sensed the same subliminal message. Even though the staircase was broken into two sections—one section that took me to the initial level, a second that took me to the top—the angles of the staircases themselves were also steep. Walking up, I couldn't anticipate what I would meet when I reached the top. I suppose that was the point—to provoke mystery and fear in the minds of all who dared see the king.

The next day, I went back up to the top and took the augmented reality tour. I learned that each level of Monk's Mound, including the top, was enclosed by a fence—as if to make sure any visitor drawing near the king was kept in fearful expectation until the last possible moment.

Back on the ground, I learned that Monk's Mound was enclosed on the ground by a 10- to 15-foot-high wooden palisade. The nearly two-mile-long defensive wall ran around the western, northern, and eastern sides of the mound, and extended south to encircle a significant chunk of the Cahokia site. I thought again of what it would have been like 800 years ago to stand outside the walls—walls that I wouldn't have been able to see over—waiting to be brought inside. I wouldn't know what would happen to me on the other side. And then, once inside, I would have had to wait on the menacing eastern side of the mound with its steep sides to the 100-foot top. If Cahokia's king wanted to intimidate all who were under him, he seemed to have picked the perfect architecture.

Monk's Mound faces a central plaza where several other mounds are arranged in a bit of a squarish shape. Walking the grounds down here, I noticed several points at which mounds were constantly in my peripheral vision. When I took the augmented reality tour the next day and saw the reconstructions of the buildings on top of the mounds, the feeling of smallness returned. Clearly, Cahokia was run by a number of people who were *up there*, and I was *down here*. This was a hierarchical society, and I had my place—on the ground level. The architecture of the central plaza repeatedly reminded me of that fact.

Yet even on the central plaza, it was impossible to avoid the sense of hierarchy among the mounds themselves. I couldn't escape the obvious supremacy of Monk's Mound over all the rest. Yes, Monk's Mound is high. But it's also a wide sprawl of a structure. The base of Monk's Mound is more than 12 acres,[36] or "about four-fifths of the area of the Pyramid of the Sun at Teotihuacán in Mexico and just a little more than the Pyramid of Khufu in Egypt."[37] It contains approximately 22 million cubic feet of fill. Standing in the central plaza and looking north, it was hard to get something other than Monk's Mound in my peripheral vision. It was a behemoth and commanded—demanded—attention. I don't see how anyone back then could have missed the message about who ultimately called the shots.

Back up on top of Monk's Mound, the view was beautiful. I could easily see the Gateway Arch and downtown St. Louis off to the west. In fact, I could see for miles to the south, west, and north. It made

36 Ibid.
37 Ibid.

me wonder if part of the mound's purpose was advance warning of approaching enemies.

What struck me most about being on top of Monk's Mound was the feeling I *didn't* have. Yes, the view was nice. But it wasn't the most magnificent sight I've ever seen. Modern skyscrapers let you see farther in the horizontal direction. I've seen more spectacular vistas even from natural formations like mountains. It wasn't the panorama that most hit me. It was the realization that I no longer had the *down here* sensation. Or to put it in terms of feelings I did have, on top of Monk's Mound *there was no one above me*. The king was high above everyone else at Cahokia. No one could stand in a place of authority over him. It made me wonder if he ever left the top to condescend to the riff-raff below.

Back on the 10-mile trail, I came across one more power flex—a flex of the darker kind. Just beyond where the southern end of the palisade would have enclosed the central plaza, stood a low, unimpressive mound. What *Mound 72* lacked in architectural shock value it made up for on the inside. Archaeologists have discovered tens of thousands of shell beads, hundreds of arrowheads, and other markers of high-status burial. But these presents were the least valuable. The most ominous evidence for power at this location was the 150 skeletons—all sacrificial victims.[38] Mound 72 provided Cahokia's citizens a daily reminder about the potential end to each of their life stories.

Mound 72 also gave new context to a quirky observation I had made back at Monk's Mound. A big brown wooden instructional sign faces the Monk's Mound parking lot. It warns the visitor of the danger of ascending Monk's Mound during a thunderstorm: Lightning strikes at the top. During the two days in which I visited Cahokia, the sky was clear and the sun was warm. I wasn't concerned about any inclement weather. But it was spring, and the wind was blowing constantly with gusts of 20-30 mph. The top of Monk's Mound provided no protection against the weather. I wondered how many times the powerful king of Cahokia had to contend with natural forces over which he had no control. I wondered how many times storms or high wind took out whoever commanded the city of Cahokia from high up in his perch. I also wondered what sort of psychological effect this might have on

38 Milner 2021, p. 120.

his subjects. How could the next king maintain power? How could he assert his authority?

I think Mound 72 supplied the answer.

A pity that the Algics, when coming face-to-face with the Talegas, didn't realize this sooner.

# 7

# Striking Back at the Empire Late A.D. 1200s

On April 9, 2024, several weeks after my trip to Cahokia, I was traveling through south-central Wisconsin to another mound site. Halfway between Milwaukee and Madison and just off of I-94 in rural Wisconsin farm country sits *Aztalan State Park*. Though contemporary with Cahokia, Aztalan was never more than a small village. Yet it holds outsized clues to its much bigger sister several hundred miles to the south.

Rolling fields on the south, west, and north along with the Crawfish River to the east form the borders. The ruins of Aztalan follow the downward slope of the western banks of the Crawfish. The two western-most platform mounds face the river.

I was struck by how insignificant the site appeared. From the northwestern corner to the southwestern corner, which is the longest axis of the rectangular plot, I walked only 1400 feet. That's the same length as the block that my house sits on back in my tiny town of 20,000 people[1] in Kentucky. Just under five football fields. Nothing like the sprawling landscape at Cahokia.

Yet the architectural stamp at Aztalan was unmistakable. The two highest (westernmost) mounds were both flat-topped. The largest, in the southwestern corner, was built into the slope. Approached from the

1 https://data.census.gov/cedsci/all?q=burlington,%20kentucky

west, it appears to have little more than a low, single level. Approached from the river side, which may have been the main means of transportation, the mound shows two levels (**Figure 41**). The base level spans 130 feet by 185 feet.[2] Walking up the eastern stairs, I remembered the same feeling of intimidation that I had walking up the stairs at Monk's Mound. It wasn't the same level of awe. But the view from atop this mound left no ambiguity to the locus of power.

A tall wooden palisade once surrounded Aztalan on all four sides, including the river-facing side. The current park contains a few spots where posts have been reconstructed. I couldn't see over the wall unless I positioned myself on higher ground outside the fort on the western side. Again, it reminded me of Cahokia.

Archaeologists describe the pottery at Aztalan as consistent with a mixed population of people. Mississippian pottery is tangled up with the pottery of whoever lived here earlier—as if missionaries from Cahokia never quite fully took over.[3] At least, that's one of the academic interpretations. Architecturally, from the palisade to the flat-topped mounds, it seems that the Cahokians called all the shots.

Aztalan never contained more than a few hundred residents.[4] I suppose it wouldn't have taken a huge contingent of Cahokian missionaries to assert themselves and take over back in the A.D. 1100s.[5] Obviously, when I visited Aztalan, no Cahokians were in charge. Their power had long since been broken. But it wasn't the Europeans who interrupted their rule.

Driving up to the site, I had seen a road sign announcing controlled burning. Around noon, when I planned to walk the 0.8-mile loop around the entire village perimeter, the burning had already begun near the river. The workers from the Department of Natural Resources let me walk the path along the western edge. On the southern side, one worker reluctantly allowed me down to the river but forbade me from going north along the water's edge through the smoke. My nostrils quickly understood why.

---

2 Birmingham and Goldstein 2005, p. 70.

3 Richards, J.D., "Aztalan and the Northern Tier of a Cahokia Hinterland," p. 115-116, 123, in: McNutt and Parish 2020.

4 Birmingham and Goldstein 2005, p. xi.

5 Birmingham and Goldstein 2005, p. 52.

About 45 minutes after my initial approach to the village from the north, I stood back in the parking lot in the west looking eastward toward the village plot. The whole area was enveloped in billowing white smoke.

When I got back to Kentucky, I found out something intriguing about my last sight picture from Aztalan. I may have inadvertently re-lived what the last Native residents had experienced. Aztalan was abandoned in the early A.D. 1200s. Archaeological evidence suggests that it was burned.[6]

◂◂ ▪ ▸▸

One hour southwest of Birmingham, Alabama, among the swarthy air so typical of the American southeast, sits a group of Mississippian mounds along a U-shaped bend in the Black Warrior River. My family and I arrived in late December 2023, three months before my trip to Cahokia, and right before the complex was to shut down for a week over Christmas. Because it was Alabama, the sunny December day meant we could still run around the mounds in short sleeves or, according to some of my children, in shorts.

Second in size only to Cahokia,[7] *Moundville* covers 325 acres and includes at least 29 mounds.[8] We followed the one-way loop by car to go from mound to mound.

The biggest mound was almost 60 feet high (**Figure 42**). I walked the steps to the top. The view over the rest of the area was commanding. Appropriate for a chieftain.

The chieftain's mound was also one of the mounds closest to the river. On the three sides of Moundville that are not exposed to the river, a sturdy palisade of tree trunks sealed with mud once stood. At regular intervals, the palisade jutted out to form guard towers.

Why leave the chieftain's mound exposed on the side without the wall? I got my answer when I went back down the steps and followed the winding wooded path north to the river. Steep drop-offs flanked

---

6 Richards, J.D., "Aztalan and the Northern Tier of a Cahokia Hinterland," p. 118, in: McNutt and Parish 2020.

7 Steponaitis, V.P. and Scarry, C.M., "New Directions in Moundville Research," p. 1, in: Steponaitis and Scarry 2016.

8 Blitz, J.H., "Moundville in the Mississippian World," p. 534, in: Pauketat 2012.

our trail. Anyone attacking from the river had plenty of natural obstacles to navigate before assaulting the chieftain's mound.

At the river, the defensive advantages of the Moundville site became even more clear. The horseshoe shape of the river reminded me of a valet drop-off. Across the river from where I was standing, the land formed a sort of peninsula in the heart of the U shape of the river. Anyone approaching by boat would have had to paddle up the sides of the U. These invaders could have been detected early. On the peninsula, a guard tower or two would have sufficed.

I realized that the chieftain had picked the safest spot.

Why such attention to defensive detail? Why were the builders of Moundville so concerned with protecting themselves? The short history of these sites suggests an answer:

> Moundville began around AD 1100 as a dispersed settlement with two small mounds. Not long after AD 1200, it experienced a burst of construction that transformed it into a major center. Most of the mounds were built at this time, as was a large, bastioned fortification wall.[9]

Moundville fell into decline soon thereafter:

> Initially, the site had a substantial resident population, which presumably provided the labor for this construction. At about AD 1300, however, the character of the site changed dramatically. Much of the resident population dispersed into the countryside, the fortifications were dismantled, and Moundville became a "necropolis," a place of ritual where the dead were brought from outlying settlements for burial.[10]

Whatever political power Moundville once had, it seems the former residents rejected it—and left for areas farther afield.

9 Steponaitis, V.P. and Scarry, C.M., "New Directions in Moundville Research," p. 1-2, in: Steponaitis and Scarry 2016.

10 Ibid., p. 2.

When I visited Cahokia in spring of 2024, it goes without saying that it was uninhabited. Whatever power it once held was gone. I could walk wherever I wanted to on the grounds and up and down Monk's Mound without needing any chieftain's or king's permission. No one manned a wooden palisade and demanded a reason for my presence. In fact, the wooden palisade had long since decayed and disappeared. The examples on site were just reconstructions.

This once-great city failed around the same time that Moundville did:

> By the start of the fourteenth century [i.e., A.D. 1300], Cahokia had fallen into noticeable decline. Its population had dropped, and places where public architecture once stood had reverted to residential use. Old mounds were even being used for ordinary burials. By then, these mounds had lost much or all of their former significance. Mound-building continued into the fourteenth century, although it had tapered off from earlier times. The site and the surrounding floodplain were largely deserted by about [A.D.] 1400.[11]

Sites nearby Cahokia experienced even more dramatic collapses. The Illinois River flows from northeastern Illinois in a southwesterly direction toward St. Louis. It merges with the Mississippi River just north of St. Louis (**Figure 56**). The Central Illinois River Valley (CIRV) residents shared in Cahokia's glory—and trauma:

> As Cahokia fragmented to the south in the late 13th century, there was an increase in the size and number of Mississippian settlements in the CIRV. … This Mississippian florescence, however, appears to have been cut short by intensification of hostilities in the region. The number of catastrophically burned and abandoned villages dating to the 13th century is a grim testament to the scale and intensity of warfare during this era.[12]

In fact, a large chunk of the region where the Mississippi and Ohio Rivers meet collapsed, forming what has been called the *Vacant Quar-*

11 Milner 2021, p. 115.

12 Wilson, G.D., "Living with War: The Impact of Chronic Violence in the Mississippian-Period Central Illinois River Valley," p. 526, 530-531, in: Pauketat 2012.

*ter* (**Figure 57**). Mound-building stopped. Ceremonial centers were no longer used. Year-round villages ceased to exist.[13] And all around the same time.

Clearly, something catastrophic happened to the Mississippian culture. Whatever it was, the first hints that the empire was beginning to crumble were felt in the early A.D. 1200s in the north (Aztalan). The effects then seemed to ripple south (Cahokia, Moundville).

What happened?

The Algics—who were massacred at the end of the last chapter—were, at Contact, comfortably on the other side of the Mississippi River, all the way over to the Atlantic (**Figure 24**). The Illinois nation covered the area where Cahokia once stood. Miami and Shawnee lived east of them. And the Wampanoag and relatives were on the east coast.

Presumably, the Algics must have escaped further massacres. But how? Apparently, they refused to take the initial massacre lying down:

> United, enraged, all declared, "To battle! Destroy them!"
>
> The [Talamatan], their northern friends, then arrived to join them.
>
> Sharp One was the sachem, the path-maker across the river.
>
> They won many victories there, driving away the Talegas.
>
> Stirring was the sachem; extremely strong were the Talegas.
>
> Breaking Open was the sachem; capturing all the great towns.
>
> The Crusher was the sachem; southward fled all the Talegas.[14]

The estimated date range for these events is A.D. 1243 to A.D. 1302—exactly in line with the archaeological data on Cahokia's collapse.

13 Meeks, S.C. and Anderson, D.G., "Drought, Subsistence Stress, and Population Dynamics: Assessing Mississippian Abandonment of the Vacant Quarter," p. 61-63, in: Wingard and Hayes 2013.

14 McCutchen 1993, p. 106, 110.

Did you catch how this remarkable turn of events came about? The Algics didn't (couldn't?) do it alone. To keep the Algic migration going eastward, the aid of the *Talamatans* was critical.

◀◀ ▪ ▶▶

Given the importance of the Talamatans to Cahokia's fall, a question naturally follows: Who were the Talamatans? In the above quote, I supplied McCutchen's transliteration of the Delaware word, rather than his English translation. He thought the allies were the *Iroquois*. However, later in the *Red Record*, McCutchen also uses *Iroquois* to translate the Delaware word *Mahongwi*. At the time the Algics engaged the Mahongwi, the main body of Algics had just come from Niagara Falls. At Contact, the area around Niagara was the homeland of Iroquoian peoples (**Figure 30**). *Mahongwi* seems like a good candidate for the *Iroquois*.

*Talamatan* does not.

Daniel Brinton, an ethnologist who published his English translation of the *Red Record* in A.D. 1885, didn't take a guess. He simply transliterated the word.

Who were the Talamatan? Who were the people who secured a victory over the greatest empire in pre-Columbian North America? A helpful place to begin our search is the description in the *Red Record* itself: They came from the *north*.

At Contact, the non-Algic peoples immediately north of Cahokia were nations like the Winnebago (Ho-Chunk) and Santee Sioux (**Figure 58**). These less famous groups were linguistic relatives of some of the more famous—and martial—peoples of the Plains: Nations like the Lakota from whom warriors like Sitting Bull, Red Cloud, and Crazy Horse hailed.

If only we had a sense for *when* the Winnebago (Ho-Chunk) and Santee arrived in Wisconsin.

◀◀ ▪ ▶▶

Linguistic relationships provide the first clues. Language comparisons don't provide the temporal element, but they trace a geographic path for Winnebago (Ho-Chunk) and Santee migrations. The path lays the foundation for other lines of evidence, and it forms a standard against which additional clues can be measured.

Winnebago (Ho-Chunk) and Santee are both members of the *Siouan-Catawban* language family. This grouping also includes members who, at Contact, were geographically far-flung. From the Crow, Stoney, and Assiniboine in the far northwest (**Figure 58**) to the Catawbans, Tutelo-Saponis, and Woccons near the Atlantic (**Figure 58**), Siouan-Catawbans almost stretched coast to coast.

The deepest linguistic split separates the Catawban languages (Catawba, Woccon) from the Siouan ones. Presumably, to form the first splinter group (the Catawbans), the splinter would have had to migrate away from the main body. At Contact, the Catawbas and Woccons were on the Atlantic. Surely the main body would have been west of them (**Figure 59**).

The next subdivision resulted in the Mandan and Missouri River Siouan (Crow, Hidatsa). At Contact, these were found on the Plains out west (**Figure 59**). Again, it seems safe to assume that the splinter groups would have had to migrate away from the main body. To reduce migration distance, it makes sense to put the main body east of the Mississippi rather than, say, west of the Great Plains. This way, the migration to form the Catawbans and the migration to form the Mandan and Missouri River Siouan (Crow, Hidatsa) groups is minimized (**Figure 59**).[15]

A central, east-of-the-Mississippi location also fits nicely with the next steps of the linguistic diversification hierarchy: The formation of Ofo, Tutelo, and the Mississippi Valley Siouan (including Lakota, Dakota, Ho-Chunk, etc.) language groups.[16] The first two of these nations were near the eastern seaboard at Contact (**Figure 59**). Mississippi

15 i.e., It would be a much longer migration if the homelands were other possible locations, like Canada or the California coast.

16 I have been following the Ethnologue classification in this section (i.e., Simons and Fennig 2018b; see also the online classification: https://www.ethnologue.com/subgroup/654/, accessed March 25, 2024). However, Campbell (1997) also has a classification hierarchy for the Siouan-Catawban languages. The Ethnologue and Campbell 1997 differ on the placement of Biloxi. (Ethnologue puts Biloxi as a separate level under "Mississippi Valley Siouan"; Campbell 1997 puts Biloxi one level higher under "Ohio Valley Siouan" and joins it into an "Ofo-Biloxi" subgroup.) Given the extreme southerly location of the Biloxi at Contact (**Figure 59**), the resolution of disagreement may alter the history of the geography of the "Mississippi Valley Siouan" dispersion.

Valley Siouan languages were primarily west of the Mississippi (**Figure 59**).

Mississippi Valley Siouan also divides into sublevels: Biloxi, Dakota (consisting of Assiniboine, Dakota, Lakota, and Stoney), Dhegihan (consisting of Kansa, Omaha-Ponca, Osage, and Quapaw), and Chiwere-Winnebago (Iowa-Oto, Ho-Chunk/Winnebago, and Missouri). Among these subgroups, their distribution alone nearly implies an origin east of the Mississippi. At Contact, their individual subgroups seemed to fan out from a central location near northern Illinois (**Figure 59**).

Together, this linguistic sequence implies an overall Siouan-Catawban homeland east of the Mississippi.

It also implies a Mississippi Valley Siouan homeland right next door to Cahokia.

◀◀ ▪ ▶▶

The indigenous histories of the Siouan-Catawban language family members are consistent with, and elaborate on, the geographic path implied by linguistics. These histories also add critical bookends to the timing of the migrations.

Geographically,

> it is said that in the long, long-ago times, the ancestral tribe of the Hidatsa and Crows once lived towards the east in the "tree country," now believed to be the western end of the Great Lakes, say south of Lake Superior and west of Lake Michigan.[17]

These homelands (**Figure 60**) are directly north of Cahokia.

> Iowas, Otos, Missouris, Omahas and Ponkas have a tradition that, at a distant epoch, they, together with the Winnebagos, came from the north; that the Winnebagos stopped on the banks of Lake Michigan while the rest, continuing their course southerly, crossed the Mississippi

17 Crow 1992, p. 16-17.

> river and occupied the places in which they were found by Europeans.[18]

"North"—and northeast—of the Contact locations for Iowa, Oto, Missouri, Omaha, and Ponca takes them across the Mississippi River and into *Wisconsin*, whose eastern shores touch Lake Michigan (**Figure 60**).

Most of these nations trace their history back even farther east:

> Ages ago the ancestors of the Omahas, Ponkas, Osages, Kansas, Kwapas, Winnebagos, Pawnee Loups (Skidi) and Rees dwelt east of the Mississippi. They were not all in one region, but they were allies, and their general course was westward. They drove other tribes before them. Five of these peoples, the Omahas, Ponkas, Osages, Kansas and Kwapas, were then together as one nation. They were called Arkansa or Alkansa by the Illinois tribes, and they dwelt near the Ohio river. At the mouth of the Ohio a separation occurred [**Figure 60**].[19]

> The Ponkas say that they had seven "old men" since they became a separate tribe. Under the fifth "old man" they first saw the pale-faces. They are now under the seventh "old man." The Omahas, according to some men of their tribe, are now under their fifth "old man." Among the Dakotas, according to some anthorities [sic], an "old man" denotes a cycle of seventy years or more. If the Ponkas use the term in this sense, and are correct in doing so, they may have had a tribal existence for about 490 or 500 years. This would extend back as far as A.D. 1390 or 1380[20] [i.e., 490 or 500 years from when the document was published in A.D. 1886].

This separation is almost a century after the dates we've observed for Cahokia's fall. But it gives us a time stamp from which we can draw deeper conclusions.

---

18 Dorsey 1886, p. 213.
19 Dorsey 1886, p. 215.
20 Dorsey 1886, p. 221-222.

Additional members of the Siouan subgroup take their origins farther east:

> The Eastern Dakota claim that the Sioux originated in the north, and came south, until, somewhere to the southeast of their starting point, they were stopped by the ocean, where they scattered and went in different directions. They fought many tribes, and finally grew stronger, and then traveled northwestward toward the prairie. When they reached Minnesota and eastern South Dakota, they came upon the Cheyenne, whom they drove out onto the prairies. The Cheyenne still remember this, according to the Dakota, and declare that their ancestors lived at Enemy Swim Lake, South Dakota. … There are Eastern Dakota now living, who claim descent from the Cheyenne who dwelt about Enemy Swim Lake.[21]

> The early history of the Mandans is involved in obscurity. … They affirm that they descend originally from the more eastern nations, near the sea-coast.[22]

> The Poncas claim that the Omahas, Osages, Kaws, and two or three other tribes, a long time ago, lived with them and spoke the same vocal language. … Big Bull, one of the headmen, said his grandfather told him that in olden times, when the above-named tribes were with them, they lived near the Atlantic Ocean, and in their westward migrations became separated. They started on this movement from near Washington, District of Columbia, the Kaws and Osages coming across to Kansas, and the Poncas and Omahas going farther north, to Northeastern Nebraska. They claimed that the Poncas were at one time where the present city of St. Louis now stands.[23]

"Stopped by the ocean" and "more eastern nations, near the sea-coast" both sound like the Atlantic. The last of the three accounts above explicitly names this body of water.

---

21 Oneroad and Skinner 2003, p. 191.

22 Maximilian, Prince of Wied, "Travels in the Interior of North America, 1832-1834," p. 254, in: Thwaites 1906.

23 Clark 1885, p. 305.

What you've just read comes from Siouan groups with deep linguistic separations. Dakota and Ponca both belong to the Mississippi Valley Siouan subcategory. Mandans were part of the earliest Siouan splits. Yet all three groups tell the same story. I think it's therefore reasonable to extrapolate their histories to all the Siouan-Catawban language family members. All must have once been part of a single group that resided along the Atlantic (**Figure 60**).

An account from the late A.D. 1800s puts a date on this part of their history:

> The Santee Sioux claim that formerly their old men kept a record of events by tying knots in a long string. By the peculiar way of tying them, and by other marks, they denoted the different events, fights, etc., and even smaller matters, such as births of children, etc. I once saw a slender pole some six feet in length, the surface of which was completely covered with small notches, and the old Indian who possessed it assured me that it had been handed down from father to son for many generations, and that these notches represented the history of his tribe for over a thousand years; in fact, went back to the time when they lived near the ocean.[24]

"A thousand years" prior to the late A.D. 1800s puts the Siouan-Catawbans on the shores of the Atlantic Ocean in the late A.D. 800s (**Figure 60**). Now we have a second bookend to hold up the chapters in the Siouan-Catawban story. By the late A.D. 1300s, a number of them—the Iowa, Oto, Missouri, Omaha, and Ponca—separated where the Ohio River empties into the Mississippi River south of Cahokia. At least 400-500 years prior, in the A.D. 800s, the whole group was near Washington, D.C., east of Cahokia. In the intervening years, the group migrated westward, with different splinter groups forming along the way (**Figure 60**).

In other words, the Siouans went east—presumably through the domain of the Talega (Mississippian) empire—during the time Cahokia rose and fell.

◂◂ ▪ ▸▸

24 Clark 1885, p. 212.

In the region to the north of Cahokia, the period leading up to, and following, the collapse at Cahokia was archaeologically dynamic. The builders of the effigy mounds ceased construction around A.D. 1150. Around the same time, the *Oneota* culture, which was distinct from *Mississippian* culture, rose and eventually dominated the former effigy mound area (**Figure 61**).[25]

At the same time that the Oneota appeared in Wisconsin, another culture arose to the south—a culture whose history draws out the final details of the epic battle that went down at Cahokia over seven centuries ago. In what is now Nebraska, northern Kansas, and even parts of northeastern Colorado and southeastern Wyoming, the *Central Plains* culture appeared.[26]

Archaeologists have associated the Central Plains culture with *Caddoan* language family members. At Contact, the *Arikara, Pawnee, Wichita, Kitsai, Yatasi*, and *Caddo* members were all west of the Mississippi (**Figure 62**). So far so good; no apparent contradictions. But also no insights into Cahokia's collapse.

The migration history of the Caddoans brings Cahokia—and the geography of the Mississippian empire—into the picture:

> There are still current among the Pawnees two traditions as to the region from which they came. ... The first of these traditions, now half forgotten, is known only to the very oldest men. It is to the effect that long ago they came from the southwest, where they used to live in stone houses. This might point to an original home for the Pawnees in Old Mexico, and even suggests a possible connection with the so-called Pueblo tribes, who still live in houses made of stone, and entered from above. Secret Pipe Chief, a very old Chau-I, the High Priest of the tribe, gave me the history of their wanderings in these words: "Long ago," he said, "very far back, all of one color were together, but something mysterious happened so that they came to speak different

25 Boszhardt, R.F., "The Effigy Mound to Oneota Revolution in the Upper Mississippi River Valley," p. 414-419, in: Pauketat 2012.

26 Ritterbush, L.W., "Configuring Late Prehistory in the Central Plains," p. 32-35, 49, in: Hill and Ritterbush 2022.

> languages.[27] They were all together, and determined that they would separate into different parties to go and get sinew. They could not all go in company, there were too many of them. They were so numerous that when they traveled, the rocks where their lodge poles dragged were worn into deep grooves. Then they were far off in the southwest, and came from beyond two ranges of mountains. When they scattered out, each party became a tribe. At that time the Pawnees and the Wichitas were together. We made that journey, and went so far east that at last we came to the Missouri River, and stopped there for a time. … There we made our fields."[28]

See **Figure 63** for a map of the migrations described above and below.

> Another very old man, Bear Chief, a Skidi, said, "Long ago we were far in the southwest, away beyond the Rio Grande. We came north, and settled near the Wichita Mountains. ... So we came from the south. After we left the Wichita Mountains, that summer we came north as far as the Arkansas River, and made our fields, and raised corn. Afterward we went to the Mississippi River where the Missouri runs into it."
>
> The second of these traditions tells of a migration from the southeast. It states that the tribe originally came from somewhere in the southeast, that is from what is now Missouri or Arkansas. They started north after sinew—to hunt buffalo—and followed up the game, until they reached the northern country—the region of the Republican and Platte rivers. … The Wichitas accompanied them part way on their journey, but turned aside when they had reached southern Kansas, and went south again.
>
> All the traditions agree that up to the time of the journey which brought the Pawnees to their homes on the Solomon, Republican, Platte and Loup rivers, the Wichitas were con-

27 For young-earth creationists, this may sound similar to the events of the Tower of Babel (Genesis 11:1-9). But I suspect this refers to a later, intra-American event.

28 Grinnell 1904, p. 223-225. I retained original spellings and conventions.

> sidered a part of the Pawnee tribe. They agree also that after this separation, the two divisions of the tribe lost sight of each other for a very long time, and that each was entirely ignorant as to what had become of the other. We know that for a long time they were at war, and the difference of the dialects spoken by these two divisions of the family shows that the period of separation was a long one.
>
> The tradition of the migration of the Pawnees from the southwest is evidently much older than the one which tells of their coming from the southeast. Most of the younger men know the latter; but for the account of the journey over the mountains from the southwest and across the Rio Grande, it is necessary to go to the very old men.
>
> It is quite possible that both stories are founded on fact; and, if this is the case, the migration from the southeast may have taken place only a few generations ago. Such a supposition would in part explain its general currency at the present time.[29]

There is archaeological evidence that the Central Plains culture expanded from southeastern locations in a north/northwesterly direction.[30] This would seem to fit the second, later part of the Pawnee migration history.

The trouble with these identifications and scenarios is the timing. According to Siouan histories, there was also a third phase to the Pawnee migrations:

> Ages ago the ancestors of the Omahas, Ponkas, Osages, Kansas, Kwapas, Winnebagos, **Pawnee** Loups (Skidi) and **Rees** [Arikara] dwelt east of the Mississippi[31] [emphasis added].

In other words, the Pawnee and Arikara were once in what became Mississippian country (**Figure 63**).

29 Grinnell 1904, p. 225-227.

30 Ritterbush, L.W., "Configuring Late Prehistory in the Central Plains," p. 33-35, in: Hill and Ritterbush 2022.

31 Dorsey 1886, p. 215.

By archaeological association with the Central Plains culture, the Pawnee and Arikara must have been west of the Mississippi by A.D. 1150—prior to the collapse of the Mississippian empire. But by the Siouan accounts, they were also once east of the Mississippi—presumably, before A.D. 1150.

In short, the Caddoans must have crossed the Mississippi before the Cahokian empire fell.

How? At first glance, the geographic extent of the Cahokian empire suggests no path around. Technically, Mississippian culture extended from Wisconsin to the Gulf coast (**Figure 41, 52**), Aztalan being one of the main manifestations of Cahokian culture in Wisconsin.

But Aztalan also holds the clues to the Caddoan dilemma. Recall from earlier that the archaeological details at Aztalan suggest that the Cahokian grip was weak at best. Cahokian artifacts were mixed with other traditions. It's as if the Cahokians had sent missionaries north, but they never fully converted, or conquered/replaced, the indigenous residents.[32]

In practice, then, peoples moving from east to west might have had a corridor of escape via northern Indiana/Illinois and southern Wisconsin (**Figure 63**). That is, if the indigenous residents of northern Indiana/Illinois and southern Wisconsin were friendly, then, yes, there may have been a corridor. The rest of the Oneota history suggests a more complex narrative.

The archaeology of Oneota indicates that they didn't stay in Wisconsin.They eventually spread into what are now Iowa, southern Minnesota, and northern Illinois. In the process, they may have displaced some of the Central Plains peoples.[33]

If so, then this would represent a movement of the Oneota *from north of Cahokia to south.*

---

32 Richards, J.D., "Aztalan and the Northern Tier of a Cahokia Hinterland," in: McNutt and Parish 2020.

33 See: Ritterbush, L.W., "Configuring Late Prehistory in the Central Plains," p. 50-51, in: Hill and Ritterbush 2022.

See also: Trabert and Hollenback 2021, p. 117, 130-131.

See also: Hollinger, R.E., "Conflict and Culture Change on the Plains: The Oneota Example," p. 270-271, in: Clark and Bamforth 2018.

Who were the Oneota? Modern archaeologists have good guesses: Iowa, Oto, Ho-Chunk, Missouri (all members the same linguistic subdivision of Mississippi Valley Siouan), and perhaps even the Ponca (member of a different linguistic subdivision of Mississippi Valley Siouan). The Santee Sioux (member of a third linguistic subdivision of Mississippi Valley Siouan) may also have been part of this complex.[34]

Oneota culture arose around A.D. 1150. It would seem, then, that the Iowa, Oto, Ho-Chunk, and Missouri were in the greater Wisconsin area by A.D. 1150. (The Caddoans may have arrived with them but continued to the Central Plains cultural locations.) If Oneota includes members of other linguistic subdivisions of Mississippi Valley Siouan, then perhaps all the Mississippi Valley Siouans were up north by this period.

Just in time—from the perspective of Cahokia's collapse—to move from north of Cahokia to south.

◂◂ ▪ ▸▸

Who were the Talamatan? Let's return to David McCutchen's explanation for translating the word as *Iroquois*. McCutchen's justification is due in part to the work of a Reverend John Heckewelder. Heckewelder documented Delaware history from the Delawares themselves in the late 1700s and early 1800s. His account identified a battle at the Mississippi and labeled the allies as *Iroquois*.[35]

McCutchen's book also includes another document added to the end of the *Red Record*. McCutchen himself does not translate the *Fragment*, which dates at least to the 1800s but supplies the translation of a John Burns. Burns identifies the *Talamatans* as *Hurons* or *Wyandots*.[36] Hurons are members of the Iroquoian language family.

The history of the Iroquois shortly before and after Contact involves a series of political alliances. The *Five Nations* (and eventually the *Six Nations*) banded together *Seneca, Cayuga, Onondaga, Oneida*, and *Mohawk* (and then eventually *Tuscarora*) into one confederacy.

34 Boszhardt, R.F., "The Effigy Mound to Oneota Revolution in the Upper Mississippi River Valley," p. 415, in: Pauketat 2012.

35 Heckewelder 1819, p. 29-32.

36 McCutchen 1993, p. 155.

McCutchen seems to think that *Talamatan* was used for one of the individual nations of the Iroquois, and *Mahongwi* for the confederacy.[37]

I suppose McCutchen could be right. The *Talamatans* may indeed have been Iroquoians, who were north of Cahokia in the late A.D. 1200s and early A.D. 1300s. At Contact, they were farther east (**Figure 30**). They could have migrated east from Wisconsin in the intervening centuries.

I suppose it's also possible that the Siouans were in Wisconsin at the time the Wyandots/Iroquois allied with the Algics. After the latter alliance defeated the Mississippian empire, the Siouans may have seen an opportunity for easy conquest. Perhaps they swooped in to take whatever pieces remained of the former Cahokian empire.

Then again, I find the archaeological correlations intriguing. They seem to imply the following narrative (**Figure 64**):

> By A.D. 1150, a century before Cahokia fell, a combined group of Siouans—perhaps even the ancestors of all Mississippi Valley Siouans—appeared in Wisconsin as the Oneota culture. In other words, the ancestors of Biloxi, Assiniboine, Dakota, Lakota, Stoney, Kansa, Omaha, Ponca, Osage, Quapaw, Iowa, Oto, Ho-Chunk/Winnebago, and Missouri were positioned north of Cahokia. Then, in the early A.D. 1200s, before Cahokia itself was toppled, the Mississippians at Aztalan abandoned their homes. Their city was burned. The geographic proximity of Oneota and Aztalan suggests a cause-effect relationship.
>
> Immediately following the decline of Cahokia, Oneota appeared to the south of their Wisconsin homelands.[38] This north-to-south movement fits the description of the Talamatans.[39]

---

37 e.g., see McCutchen 1993, p. 112, 129, 131.

38 Trabert and Hollenback 2021, p. 130-131.

39 It also may conveniently explain the origin of the linguistic subgroups. In other words, the march south may have been the catalyst for the formation of the Biloxi who, at Contact, were much farther to the south on the Gulf (**Figure 59**); and for the formation of the Kansa, Omaha, Ponca, Osage, Quapaw, Iowa, Oto, and Missouri, who may have continued farther south past Cahokia to "the mouth of the Ohio" where subsequently "a separation occurred." Presumably, the Winnebago/Ho-Chunk remained in Wisconsin, and the Dakota may have simply migrated west.

> None of these southward-migrating groups seem to have moved passively and peacefully. The violence at the Central Illinois River Valley that we discussed at the beginning of this chapter was around this time. It was also south of Wisconsin and presumably in the path of the Siouan marches.[40]

We may never know whether the Talamatans were Siouans or Iroquoians. Regardless, we do know the effects of their alliance with the Algics. Together, they inflicted a catastrophic defeat on the Cahokia empire—one from which the Mississippians would never fully recover.

40 Hollinger, R.E., "Conflict and Culture Change on the Plains: The Oneota Example," p. 269, in: Clark and Bamforth 2018.

# 8

# Dénouement
# A.D. 1300s to Contact

The Contact map of the eastern United States shows Algonquians spread all over the eastern midwest, the northeast, and the Atlantic coast (**Figure 65**). Few other language groupings fill the scene. Muskogeans occupy the southeast (**Figure 65**). But these may have been recent immigrants, refugees fleeing the invasions of the Conquistadors in Mexico (see **chapter 6**). The eastern Siouan-Catawbans (Woccons, Tutelo-Suponis, etc.) (**Figure 65**), on the other hand, must have been established before the Algonquians marched in (see **chapter** 7). Iroquoians, the last major Contact-era language grouping in the east, split the Algonquian tribes in two (**Figure 65**).

After defeating the Talegas at the Mississippi River, the Algonquians would never face as severe a test again. At least, not until European Contact. Between these two bookends, though, the Algonquians had much to accomplish.

It is to these conflicts, and to the final developments that gave rise to the Contact-era map, that we turn next.

◂◂ ▪ ▸▸

Once the Algics defeated the Natchez, the Algonquian allies, the Talamatan (see **chapter** 7) turned on them (estimated date range of A.D. 1280 to A.D. 1332):

> South of the Great Lakes we lit our fires; north of the Great Lakes were [Talamatanitis].
>
> But they were not true friends and conspired; Long in the Wood was the sachem.[1]

The Algonquians then freed themselves of the burden of their former allies:

> Truthful Man was the sachem who fought against the [Talamatan].
>
> Righteous was the sachem; trembling were the [Talamatan].[2]

After this victory, the Algonquians east of the Mississippi enjoyed 50 to 100 years of peace. Based on geographic clues in the *Red Record*, their residence was somewhere in the Indiana/Ohio/Kentucky area. The *Red Record* implies significant population growth during this period.[3]

This time frame also saw the next population split in the Algonquian group: The Core Central Algonquian language branch formed. The *Red Record* describes the initial event for one of the members of the Core Central Algonquian branch, the Shawnee (estimated time frame for this event is A.D. 1362 to 1391):

> Little Cloud was the sachem; many were those who left.
>
> The Nanticokes and the Shawnee went to the south land.[4]

Around the same time, an individual called the "Prophet"[5] also left the main body and headed back to visit earlier splinter groups. Nothing in the *Red Record* implies internal conflict as a cause for the migrations. Presumably, the "great harvests"[6] and population growth of this period led to friendly goodbyes.

About two centuries later (i.e., around A.D. 1554), the Shawnee had left "the south land" and were back up in their old stomping grounds:

1 McCutchen 1993, p. 110
2 McCutchen 1993, p. 110
3 McCutchen 1993, p. 116
4 McCutchen 1993, p. 116.
5 McCutchen 1993, p. 120
6 McCutchen 1993, p. 116

> White Horn was the sachem west to the Ohio.
>
> There, were the Illinois; there, were the Shawnee; there, were the Conoy.[7]

The Shawnee's own history agrees with the *Red Record*:

> John Johnston, Esq., who was for many years [the Shawnee] agent, prior to 1820, observes...that they migrated from West Florida, and parts adjacent, to Ohio and Indiana, where this tribe was then located.[8]

What were the Shawnee doing as far south as Florida? The dates for their absence from the Indiana/Ohio/Kentucky area (the mid- to late-1300s to the mid-1500s) overlap in time with political change observed among the Mississippians in the southeast:

> Individual Mississippian societies emerged and declined across the landscape, rarely existing for more than a century or two, with the expansion of one typically at the expense of others elsewhere in the surrounding region, in processes variously described as cycling or fission-fusion. ... Regional-scale maps...show these societies appearing and disappearing in a pattern comparable to the blinking lights on a Christmas tree.[9]

Perhaps the Shawnee were instrumental in precipitating this turbulent period among Mississippian cultures.

◂◂ ▪ ▸▸

The split that gave rise to the Shawnee surely was the split that also gave rise—eventually—to their linguistic relatives: The Fox, Sac, Kickapoo, Mascouten, Miami-Illinois, Ojibwa-Potawatomi-Ottawa, Algonquin (Algonkin), and Saulteaux peoples. One of these groups, the Ojibwa ("Ojibway"), possesses a detailed migration account, as per one of their priests, who told it to a Minnesotan named William Warren.[10]

---

7 McCutchen 1993, p. 132.
8 Schoolcraft 1851, p. 19.
9 Anderson and Sassaman 2012, p. 166.
10 Warren 1885.

Apparently, the Ojibwa went far afield from the Indiana/Ohio/ Kentucky area where the main body of Algonquians remained. According to the Ojibwa priest:

> Our forefathers, many strings of lives ago, lived on the shores of the Great Salt Water in the east. ... Our forefathers moved from the shores of the great water, and proceeded westward. ... The Me-da-we lodge [i.e., religious house] was pulled down and it was not again erected, till our forefathers again took a stand on the shores of the great river near where Mo-ne-aung (Montreal) now stands.[11]

At Contact, the Ojibwa were located by the Great Lakes (**Figure 24**), which contain fresh water, not salt water. Therefore, the Great Lakes must not have been the "Great Salt Water in the east."

Modern Montreal sits along the St. Lawrence River, which empties eastward into the Atlantic Ocean (**Figure 66**). Naturally, the "Great Salt Water in the east" must have been the Atlantic Ocean.

The Ojibwa migrations continued west from Montreal:

> In the course of time this town was again deserted, and our forefathers still proceeding westward, lit not their fires till they reached the shores of Lake Huron, where again the rites of the Me-da-we were practised.
>
> Again these rites were forgotten, and the Me-da-we lodge was not built till the Ojibways found themselves congregated at Bow-e-ting (outlet of Lake Superior), where it remained for many winters. Still the Ojibways moved westward, and for the last time the Me-da-wa lodge was erected on the Island of La Pointe, and here, long before the pale faces appeared among them, it was practised in its purest and most original form.[12]

Warren related the rest of the history from La Pointe:

> The common class of the tribe who are spread in numerous villages north and west of Lake Superior, when asked where they originally came from, make answer that they

11 Warren 1885, p. 79-80.
12 Warren 1885, p. 79-80.

# Figure Section

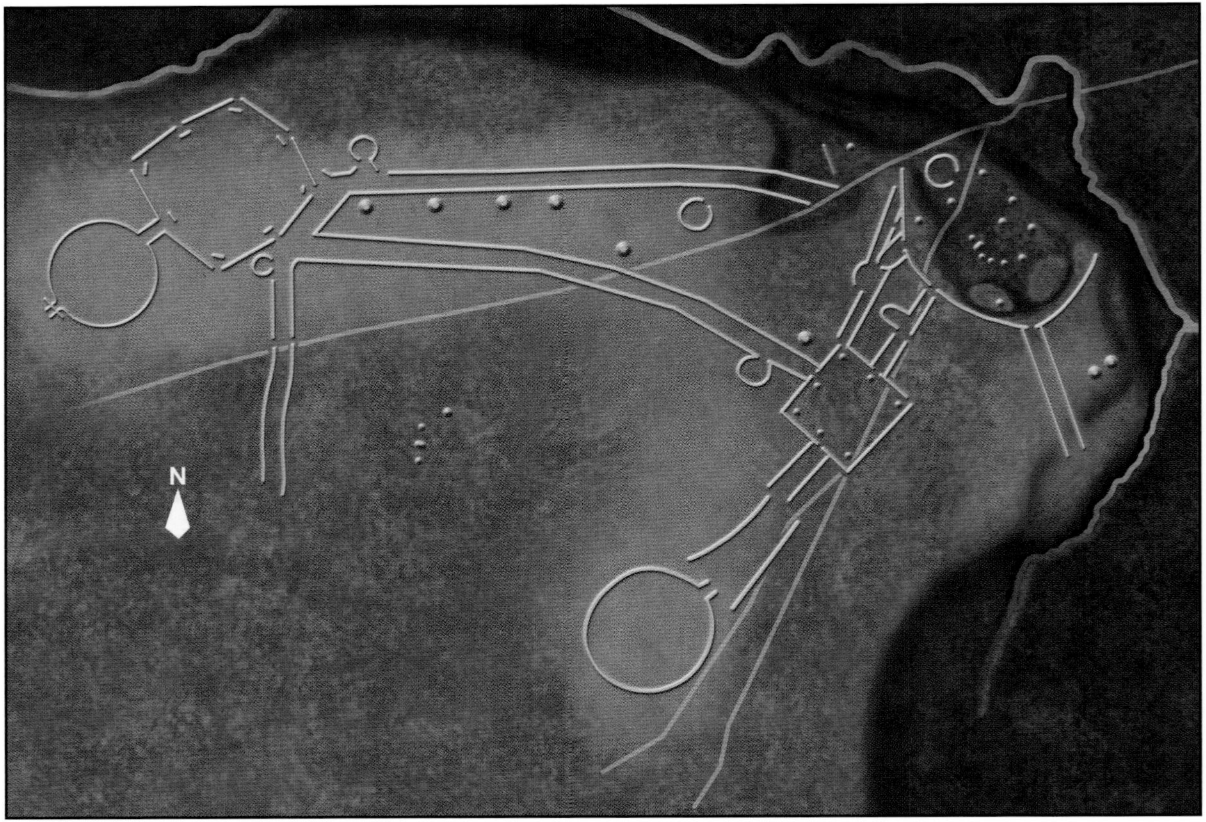

**Figure 1:** Redrawn after Squier and Davis (1848) plate number XXV. Depiction of earthworks in Newark, Ohio. The Great Circle earthwork can be found near the bottom-center of the image (eyeball-like structure). Semi-parallel lines (representing earthen walls) extend from the Great Circle to a square (another earthwork) that sits slightly off center in the image and is to the right and above the circle on the map. Other parallel lines on the map represent earthen walls. One set connects the square to the circle and octagon in the upper left of the map.

**Figure 2:** Ditch inside the Great Circle earthwork at Newark, Ohio. The circle wall rises on the right. The flat center of the enclosed area can be seen on the left.

**Figure 3**: Entrance to the Great Circle. The walls of the circle rise on the left and right. Perspective of the viewer is facing west/southwest.

**Figure 4**: Circle, causeway, and octagon earthworks at Newark, Ohio. These earthworks are now part of a golf course. The tee (center of the image) for one of the holes sits in the causeway. The causeway extends to the left and down where it links to the circle (not visible in this image). The other side of the causeway links to the octagon, for which part of one of the walls is visible on the left of the image. The earthen rise behind the tee is a small stand-alone wall within the octagon. (See **Figure 1** for a map of the octagon and the stand-alone structures within it.)

**Figure 5**: Simulation of the lunar rise over the Newark, Ohio, circle-and-octagon earthwork. (See **Figure 1** for the geographic relationship between these earthwork structures and the others at Newark.)

**Figure 6**: Earthwork mounds from Mound City Group in Chillicothe, Ohio.

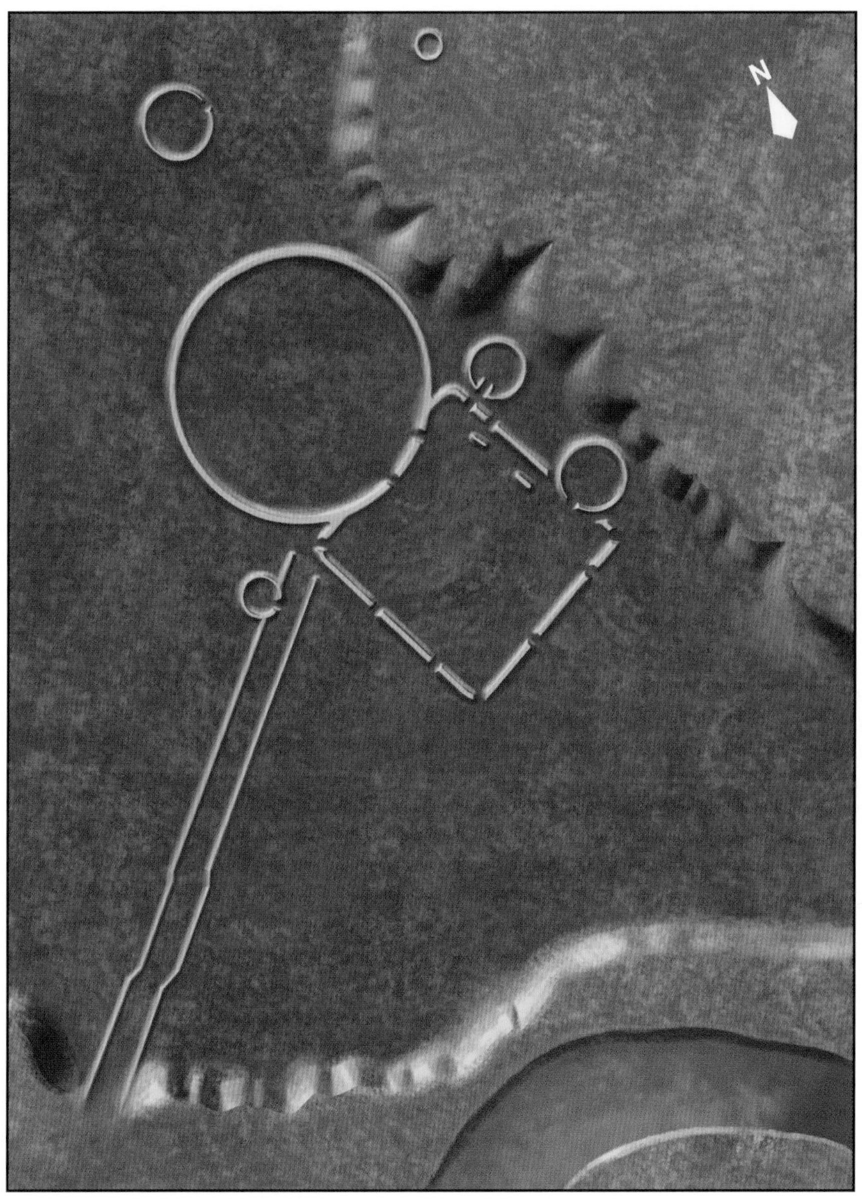


**Figure 7**: Map of Hopeton earthworks in Chillicothe, Ohio. Redrawn after Squier and Davis (1848) plate number XVII.

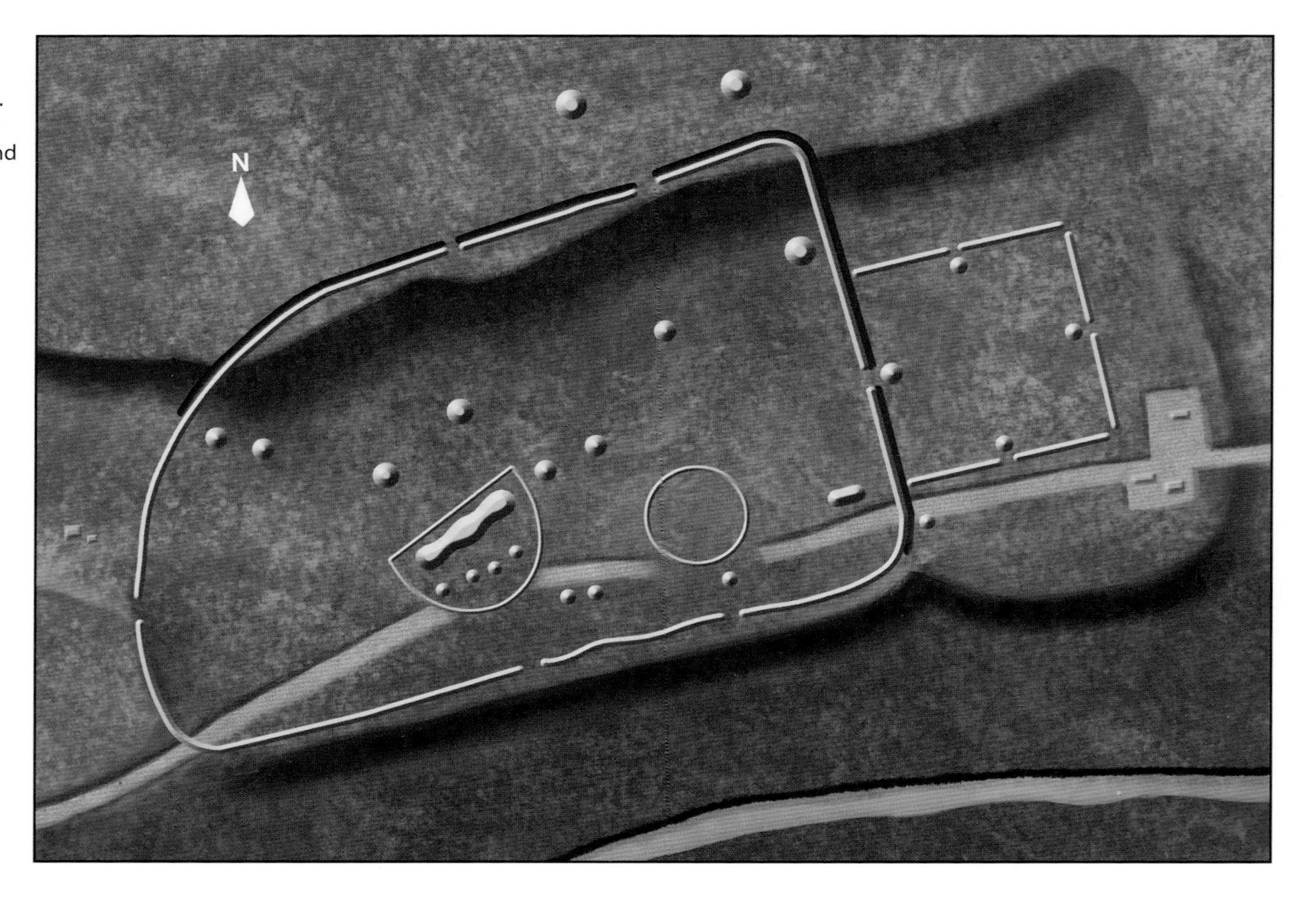


**Figure 8**: Map of Hopewell Mound Group. Redrawn after Moorehead and Laufer (1922) opening fold-out map.

**Figure 9:** Far-flung economic reach of the Hopewell culture. Lines trace to geographic origins of materials found at Ohio Hopewell sites.

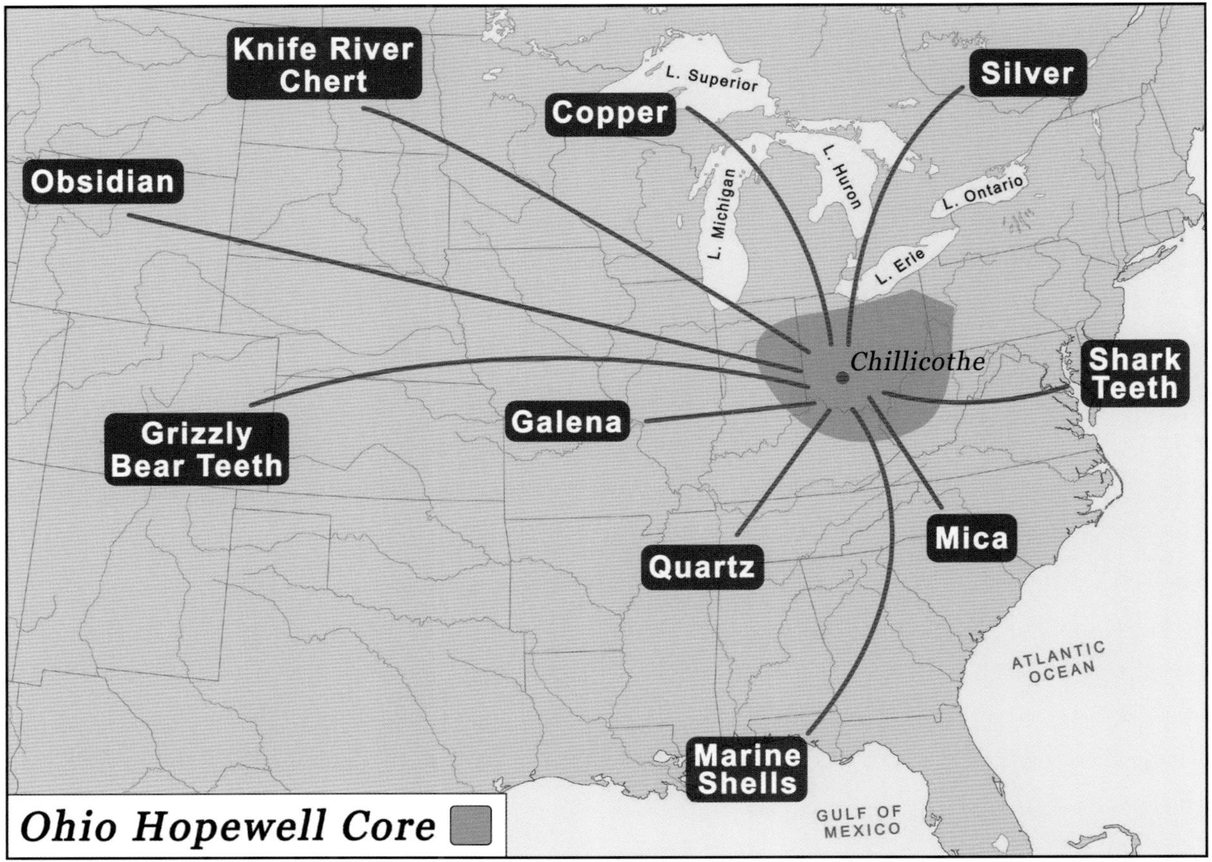

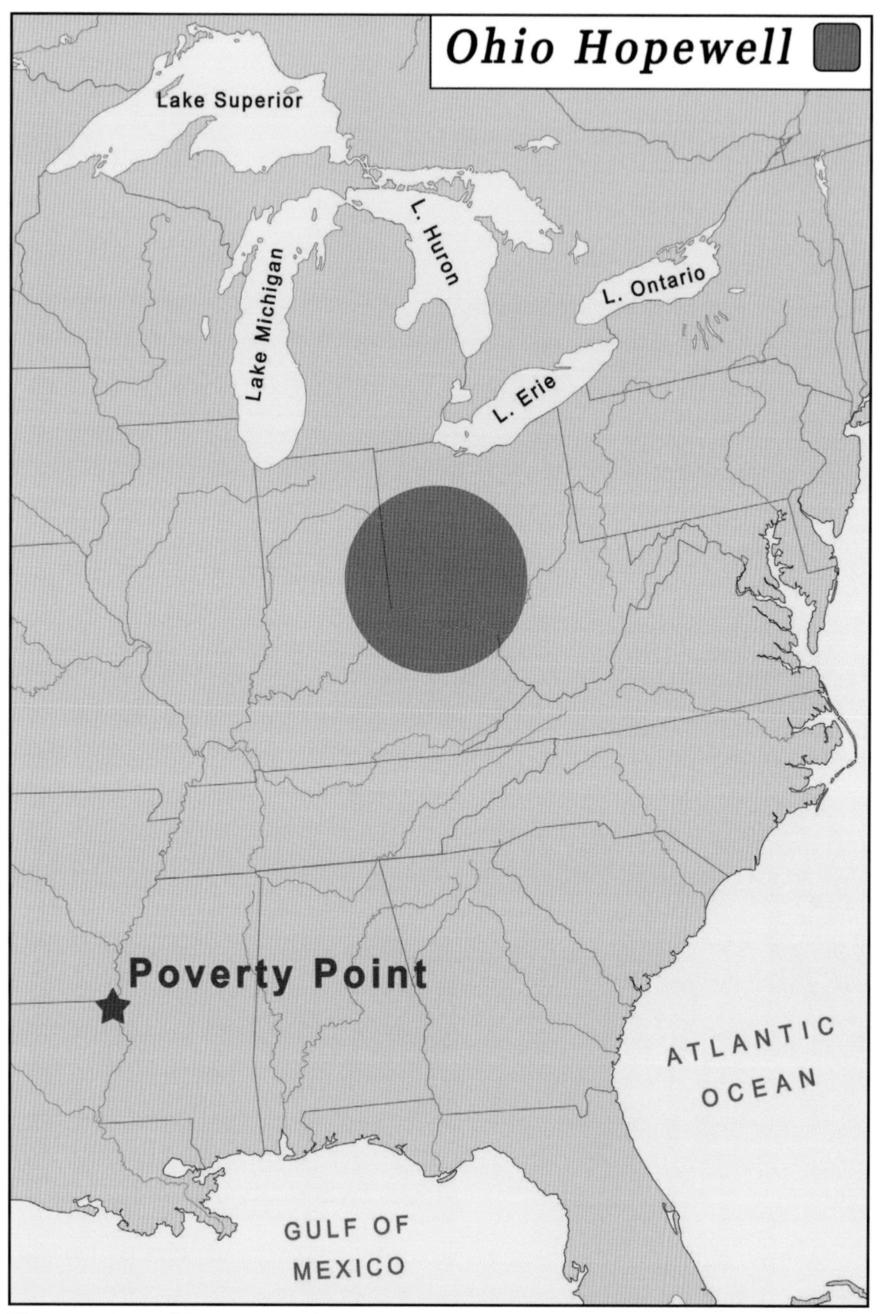


**Figure 10**: Map of Poverty Point location as well as core area of Ohio Hopewell.

**Figure 11**: Reconstruction of Poverty Point, the ruins of which are located in what is now northeast Louisiana.

**Figure 12**: Giant sculpted head of Olmec origin.

**Figure 13**: Location of Olmec heartlands. Just across the Gulf of Mexico was Poverty Point in Louisiana.

**Figure 14**: Olmec earthen mound at La Venta, Tabasco, Mexico.

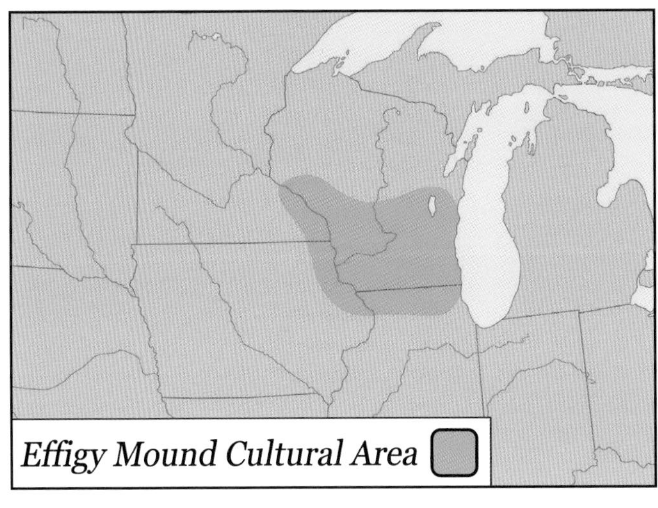


**Figure 15**: Homelands of the Effigy Mound culture.

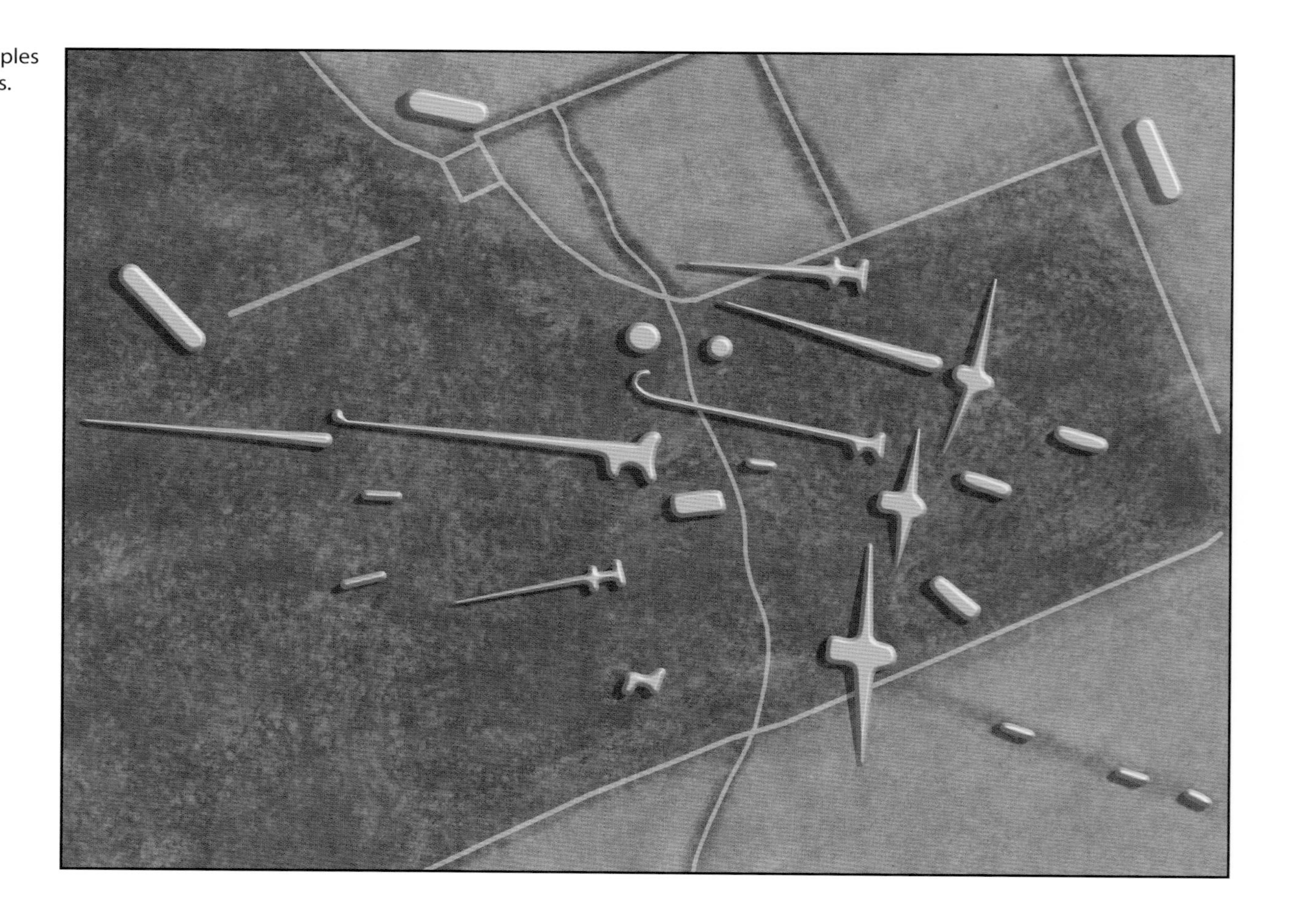

**Figure 16:** Examples of effigy mounds. Redrawn after Lapham (1855) plate XVI.

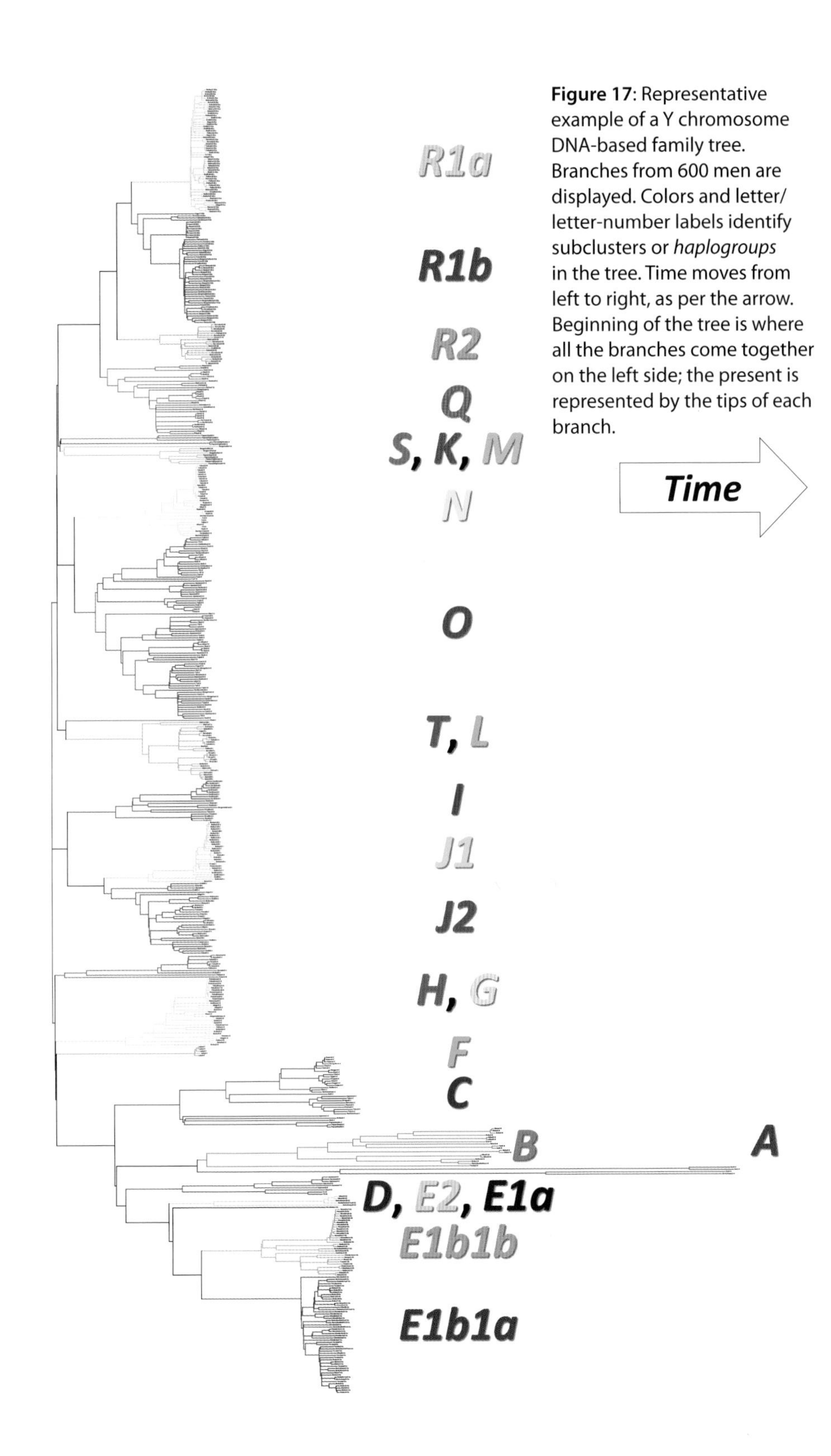


**Figure 17**: Representative example of a Y chromosome DNA-based family tree. Branches from 600 men are displayed. Colors and letter/letter-number labels identify subclusters or *haplogroups* in the tree. Time moves from left to right, as per the arrow. Beginning of the tree is where all the branches come together on the left side; the present is represented by the tips of each branch.

**Figure 18**: Global distribution of Y chromosome DNA haplogroup (branch) *Q*. Size the of the circle represents relative abundance. Frequencies below 1% are not displayed on the map.

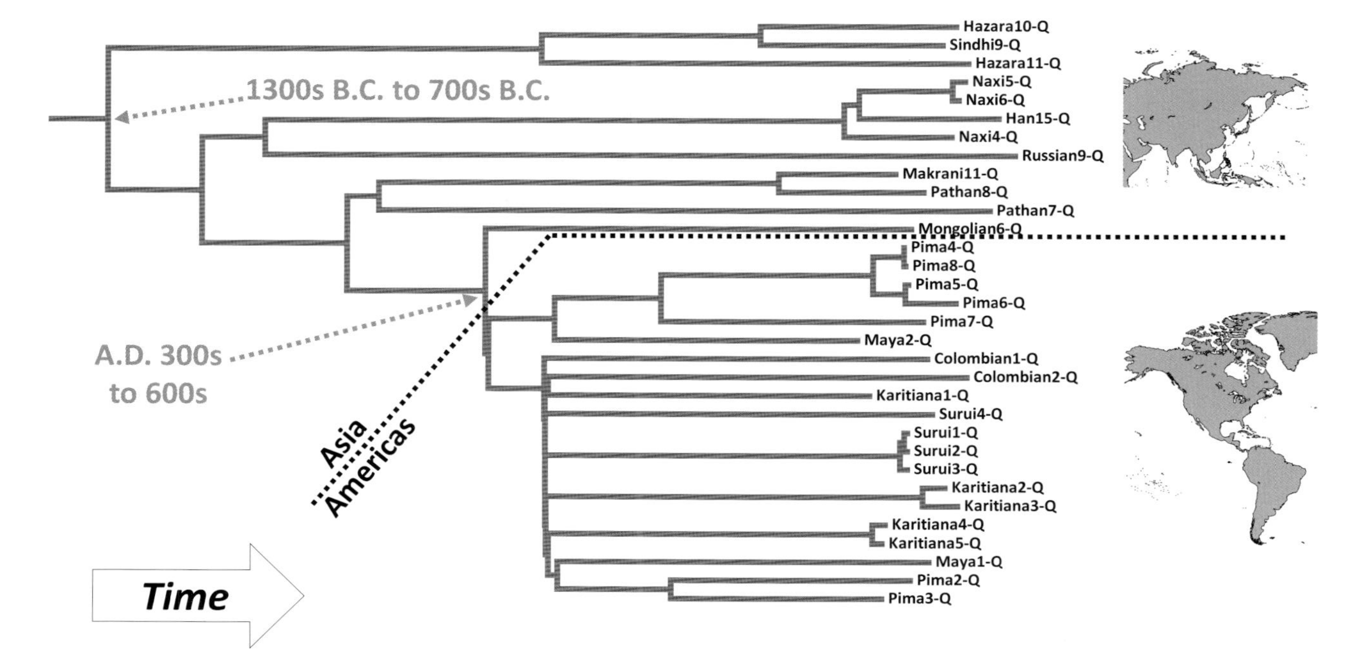


**Figure 19**: Zoomed-in view of haplogroup (branch) *Q* from **Figure 17**. American men include the *Pima* from northwest Mexico, the *Maya* from the Mexican Yucatan, *Colombians* from Colombia, and the *Karitiana* and *Surui* from Brazil. Eurasian men include *Hazara* from Afghanistan, *Sindhi*, *Makrani*, and *Pathan* from Pakistan, *Naxi* from southern China, *Han* from China, *Russians* from western Russia, and *Mongolians* from Mongolia. Date ranges indicate the calculated time of branch separations. The dashed black line identifies the place in the tree where the American and Asian branches separate. Numbers to the right of the ethnic labels are simply arbitrarily assigned to differentiate individuals.

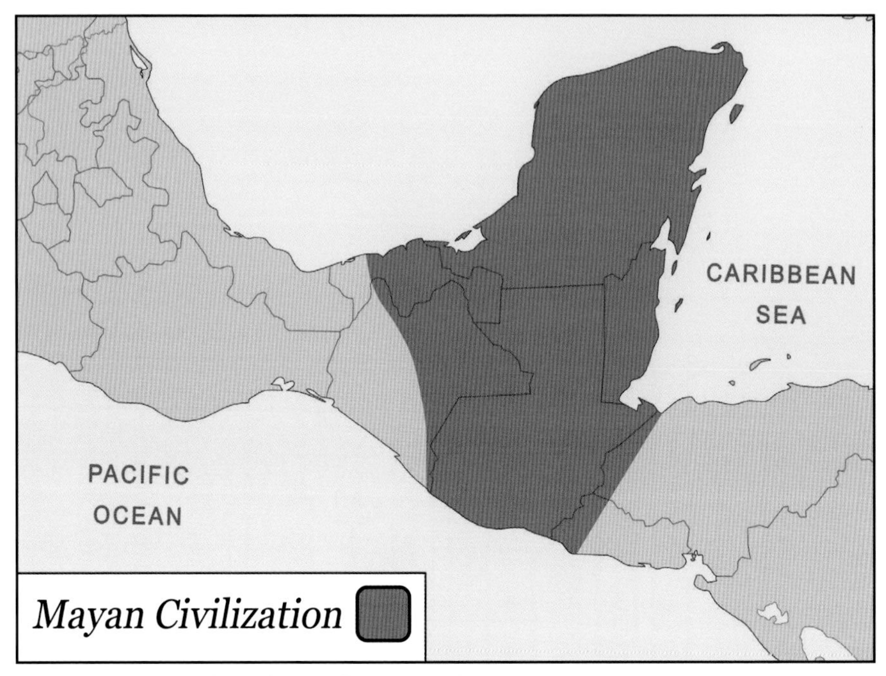


**Figure 20**: Ancestral homelands of Maya peoples.

**Figure 21**: Example of Mayan writing.

**Figure 22**: Teotihuacan, Mexico. Pyramid of the sun is the largest building behind the smaller structures in the foreground.

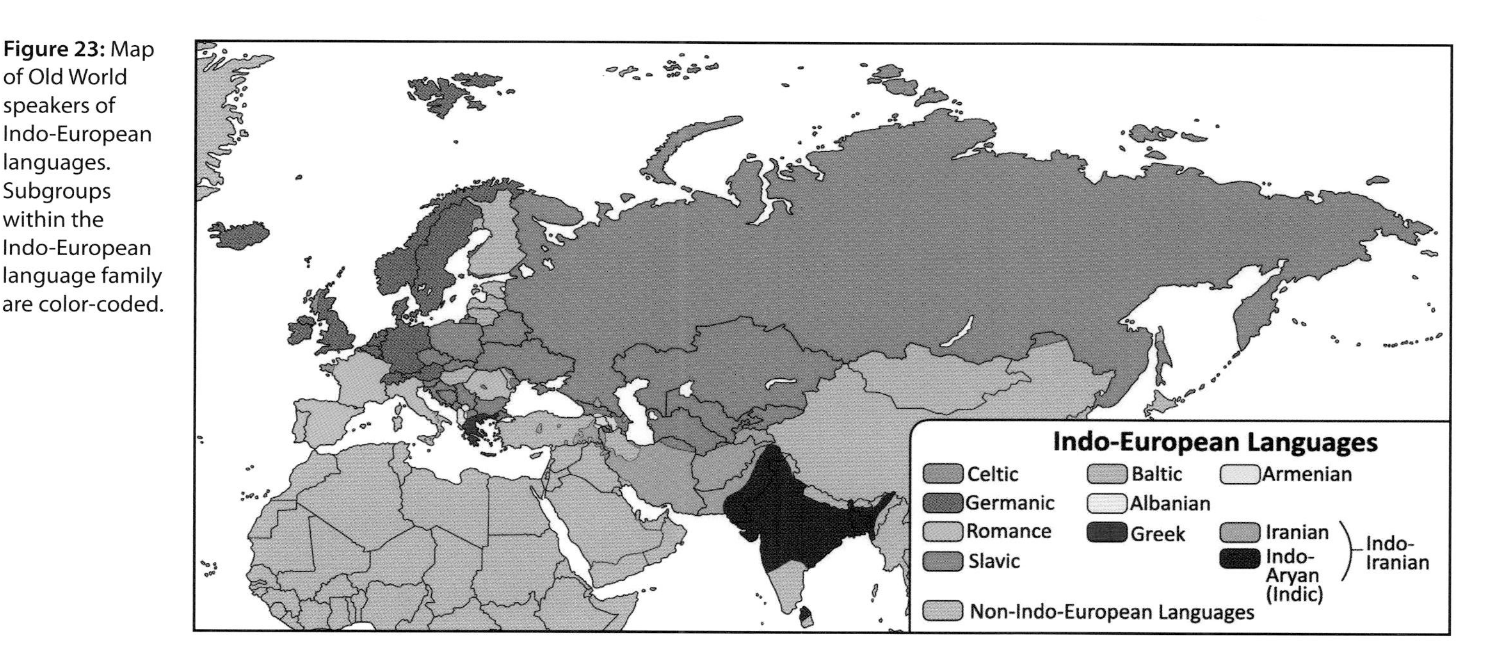


**Figure 23:** Map of Old World speakers of Indo-European languages. Subgroups within the Indo-European language family are color-coded.

**Figure 24:** Contact-era geographic distribution of members of the Algic language family.

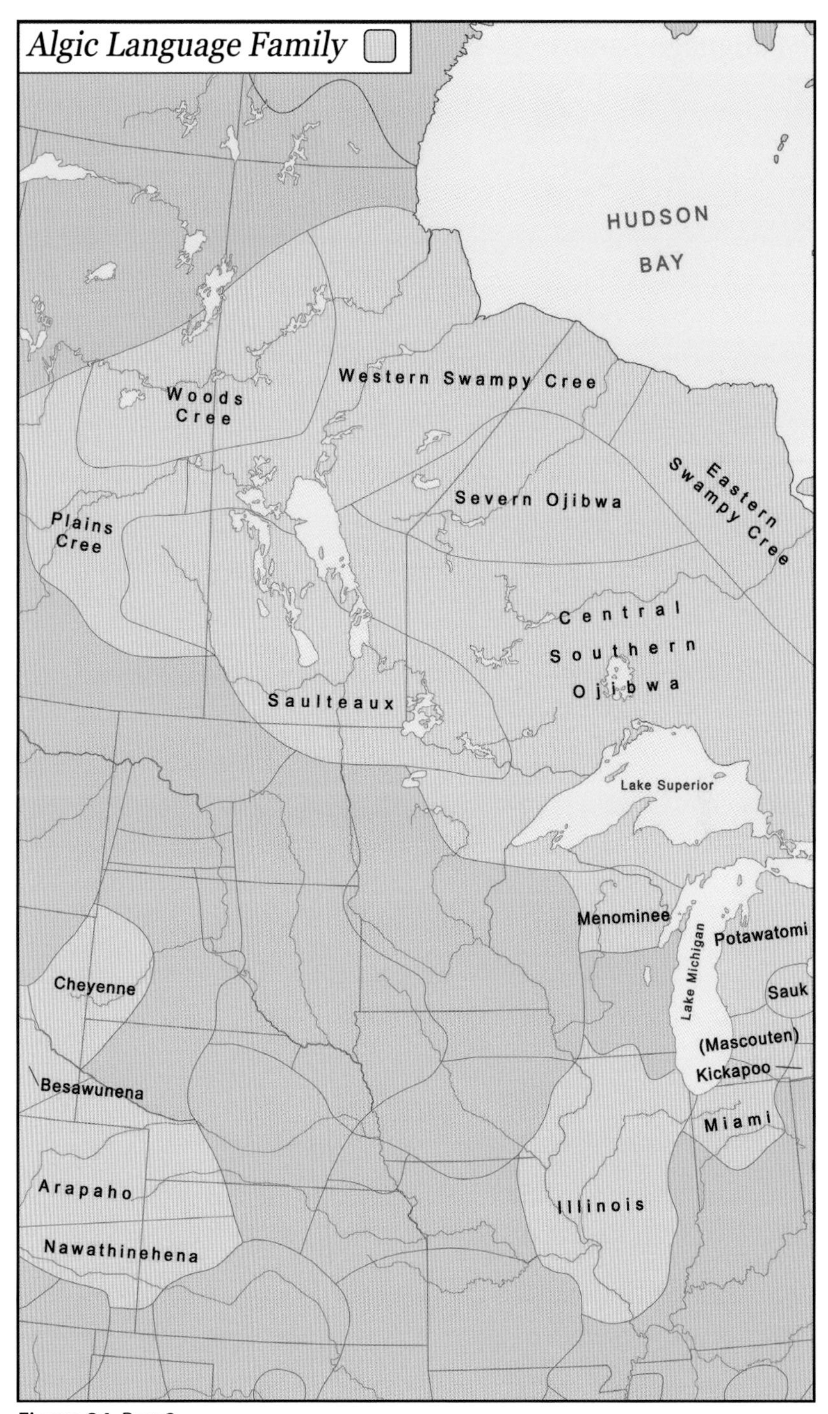


**Figure 24**: Part 2

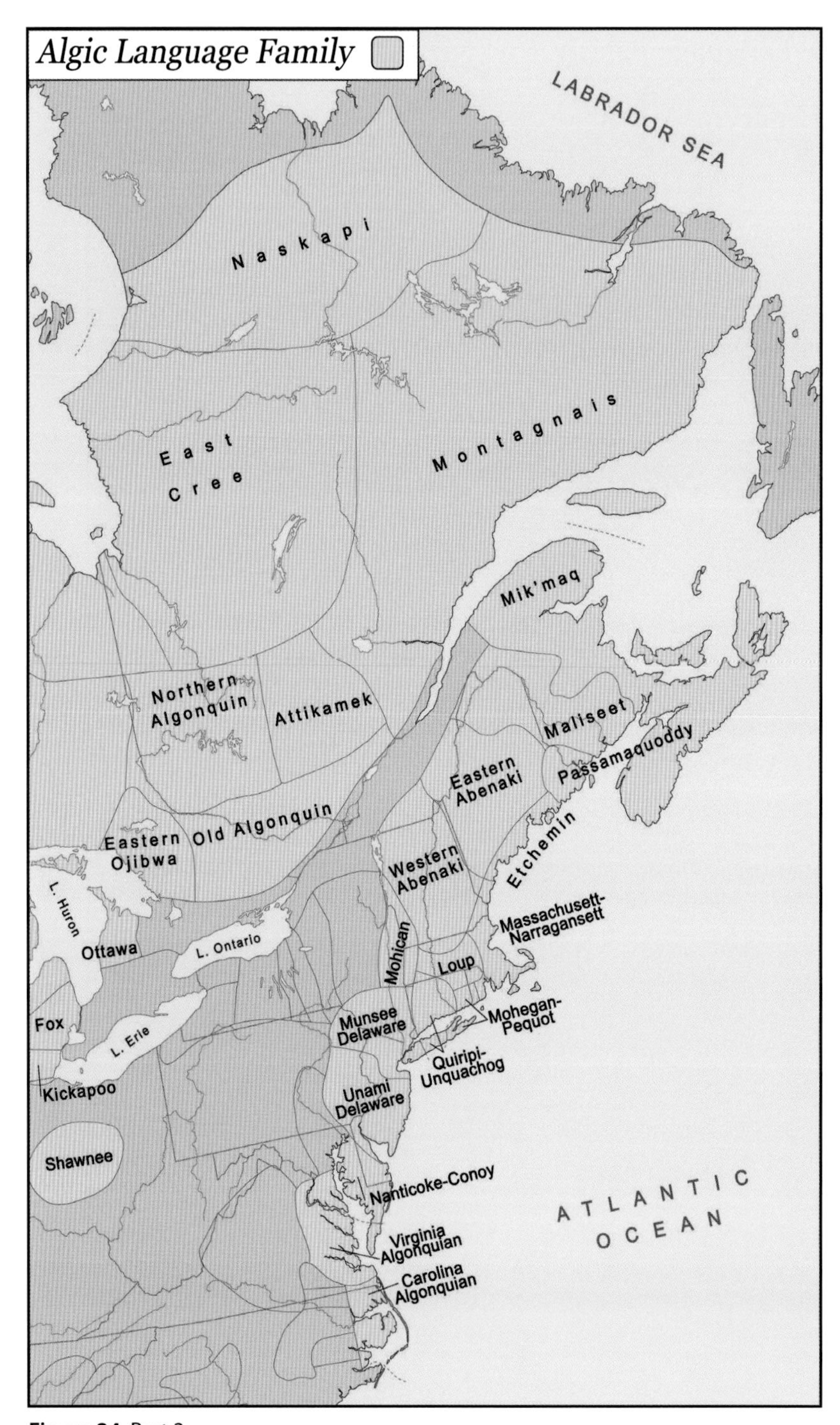


**Figure 24**: Part 3

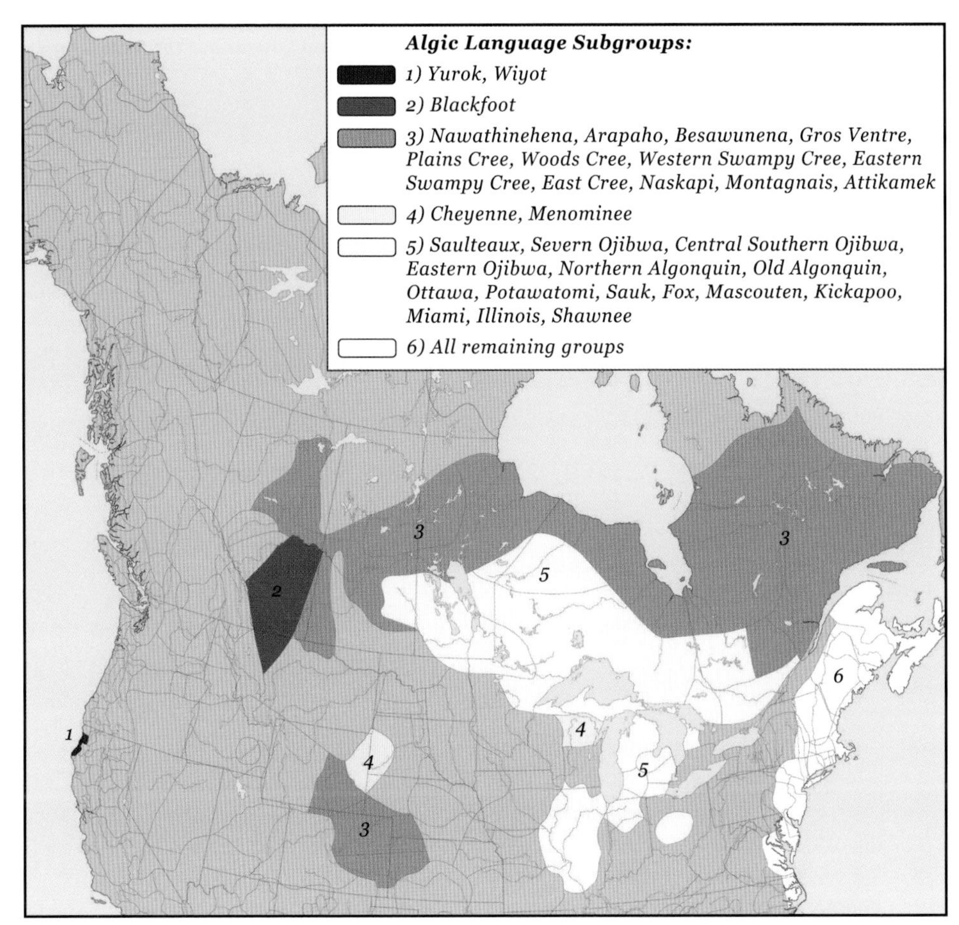


**Figure 25:** Contact-era geographic distribution of members of the Algic language family, now colored according to linguistic subgroup. The oldest linguistic subgroup (1) is in the west; the youngest (6), in the east.

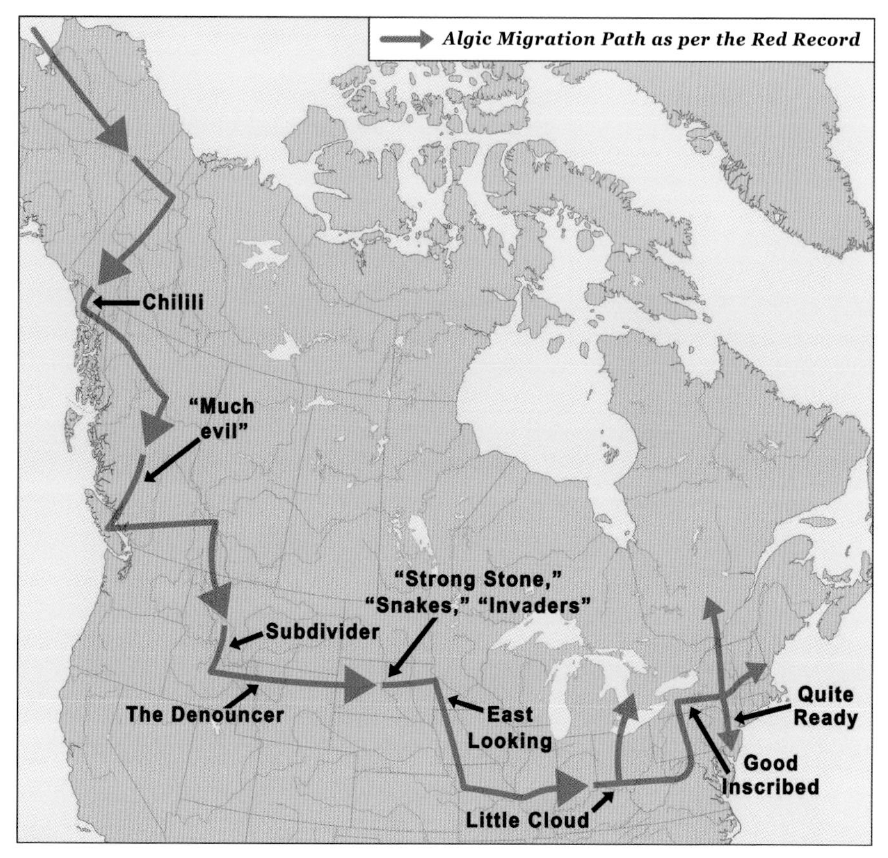


**Figure 26**: Migration of the Algic as per the McCutchen's (1993) reconstruction from geographic details in the *Red Record*.

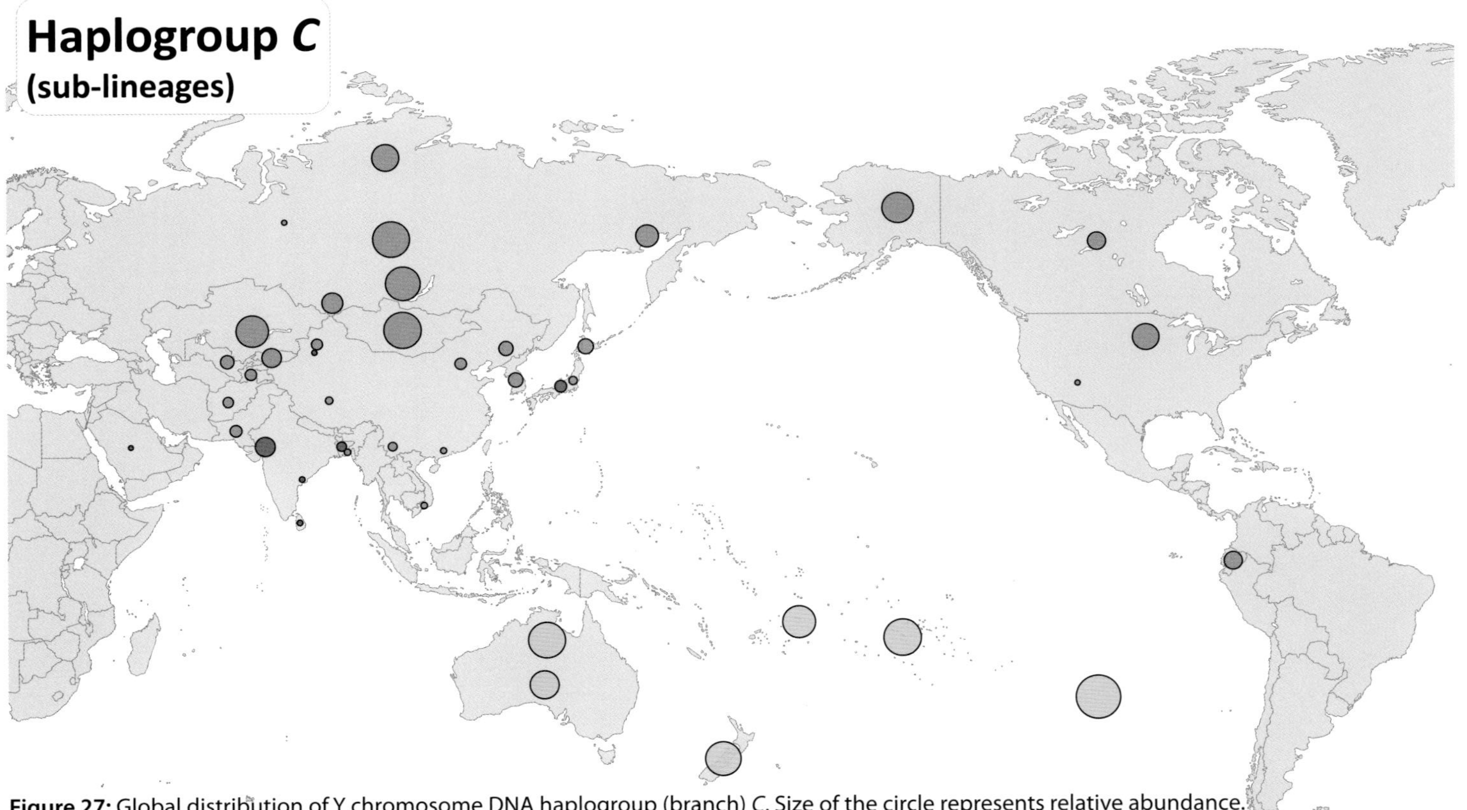


**Figure 27:** Global distribution of Y chromosome DNA haplogroup (branch) *C*. Size of the circle represents relative abundance. Frequencies below 1% are not displayed on the map. Subbranches within haplogroup *C* have been color-coded. The American haplogroup *C* branches show the closest affinity to the Siberian/northern Eurasian subbranches.

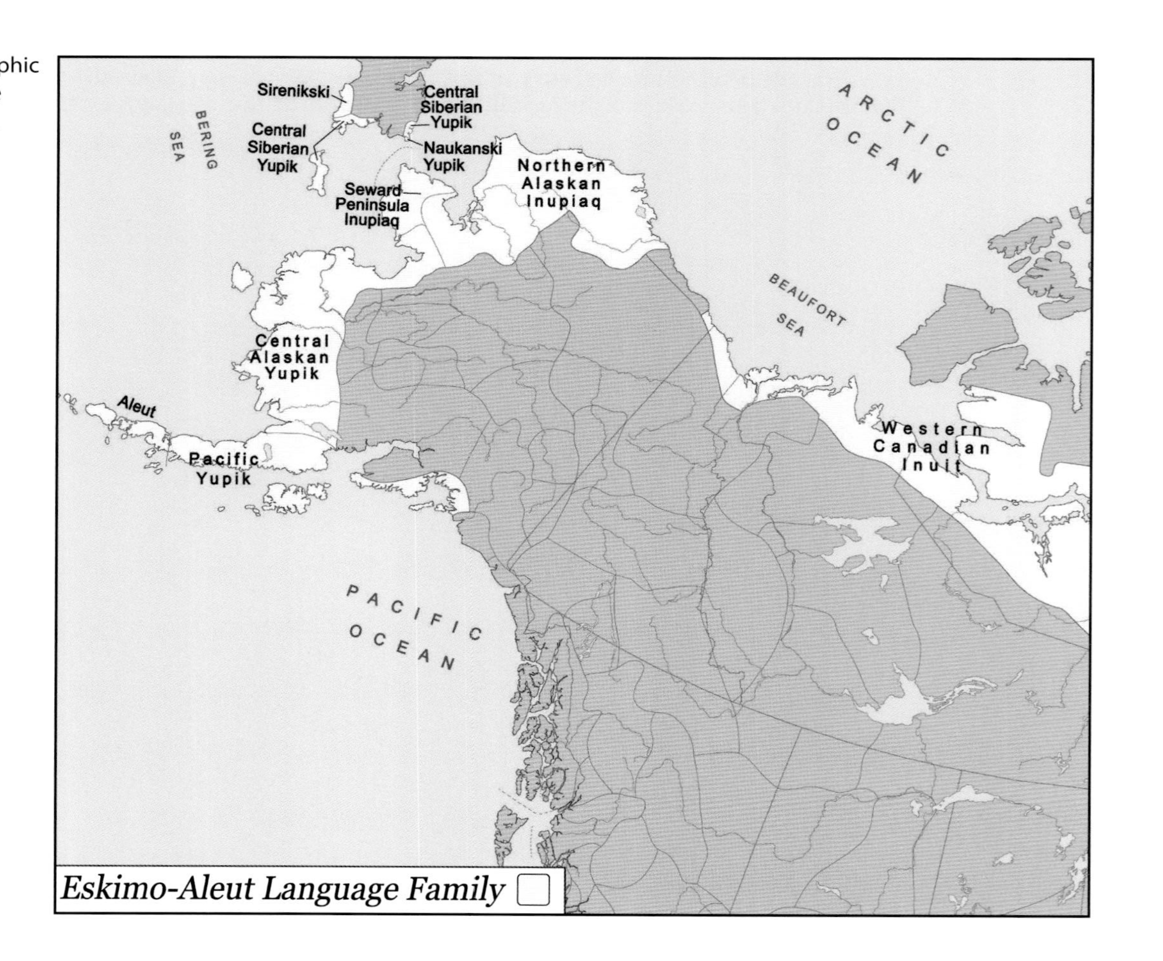


**Figure 28**: Contact-era geographic distribution of members of the Eskimo-Aleut language family.

**Figure 28**: B

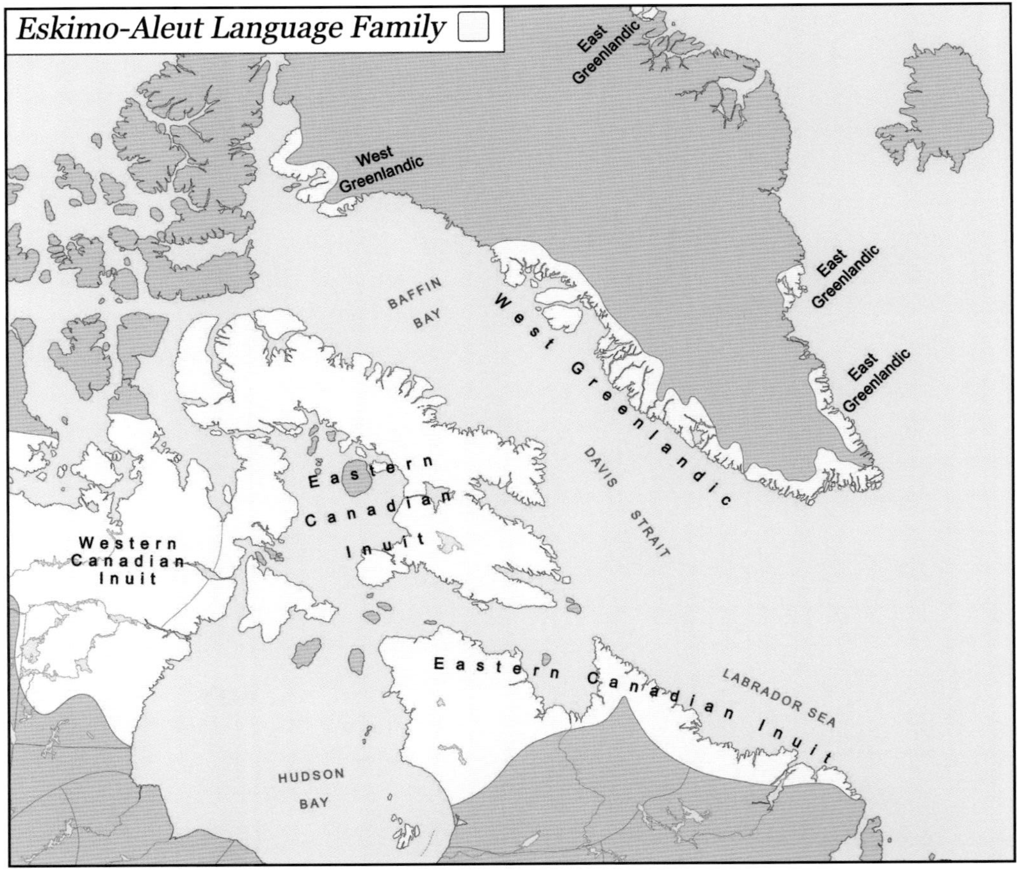

*Eskimo-Aleut Language Family*

East Greenlandic
West Greenlandic
East Greenlandic
East Greenlandic
BAFFIN BAY
West Greenlandic
DAVIS STRAIT
Eastern Canadian Inuit
Western Canadian Inuit
Eastern Canadian Inuit
LABRADOR SEA
HUDSON BAY

**Figure 29:** Contact-era geographic distribution of members of the Eyak-Athabaskan language family.

**Figure 29:** B

Athabaskan Language Family
Kwalhioqua
Clatskanie
Upper Umpqua
Tututni
Galice
Applegate
Tolowa
Hupa
Mattole
Eel River
Cahto
Navajo
Jicarilla
Western Apache
Mescalero
Chiricahua
Kiowa Apache
Lipan
PACIFIC OCEAN

**Figure 30:** Contact-era geographic distribution of members of the Eyak-Athabaskan and Iroquoian language families.

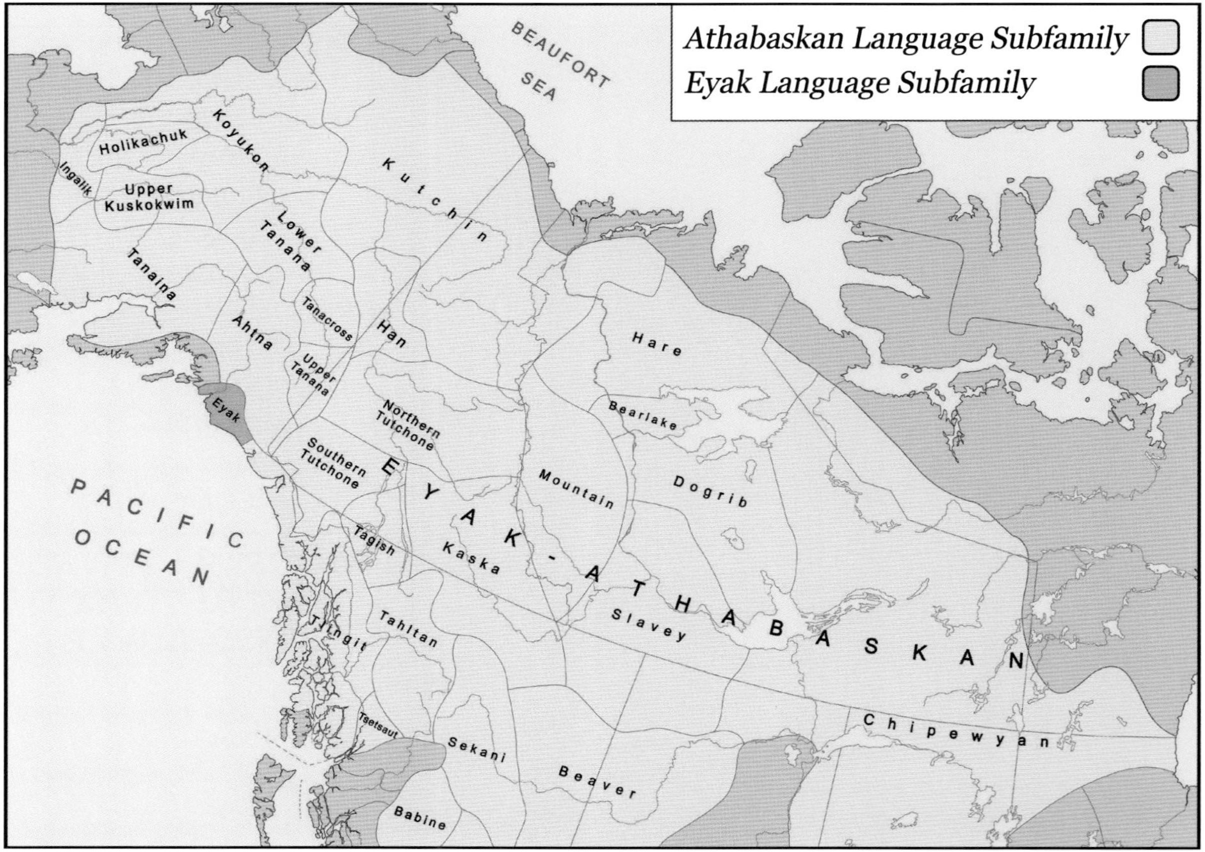

**Figure 31:** The deepest linguistic split among members of the Eyak-Athabaskan language family was between the Eyak and Athabaskan languages. The map displays Contact-era geographic distribution of the northern members of the Eyak-Athabaskan language family.

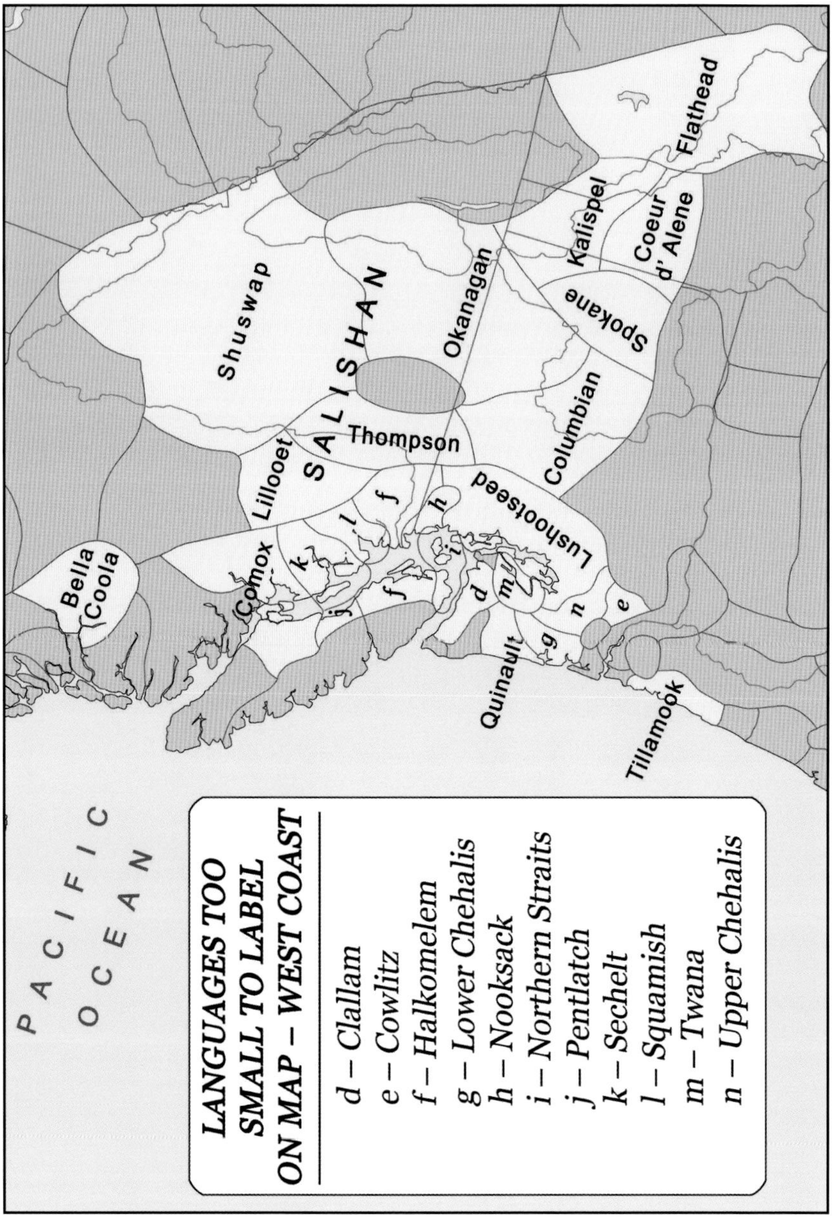

**Figure 32:** Contact-era geographic distribution of members of the Salish language family.

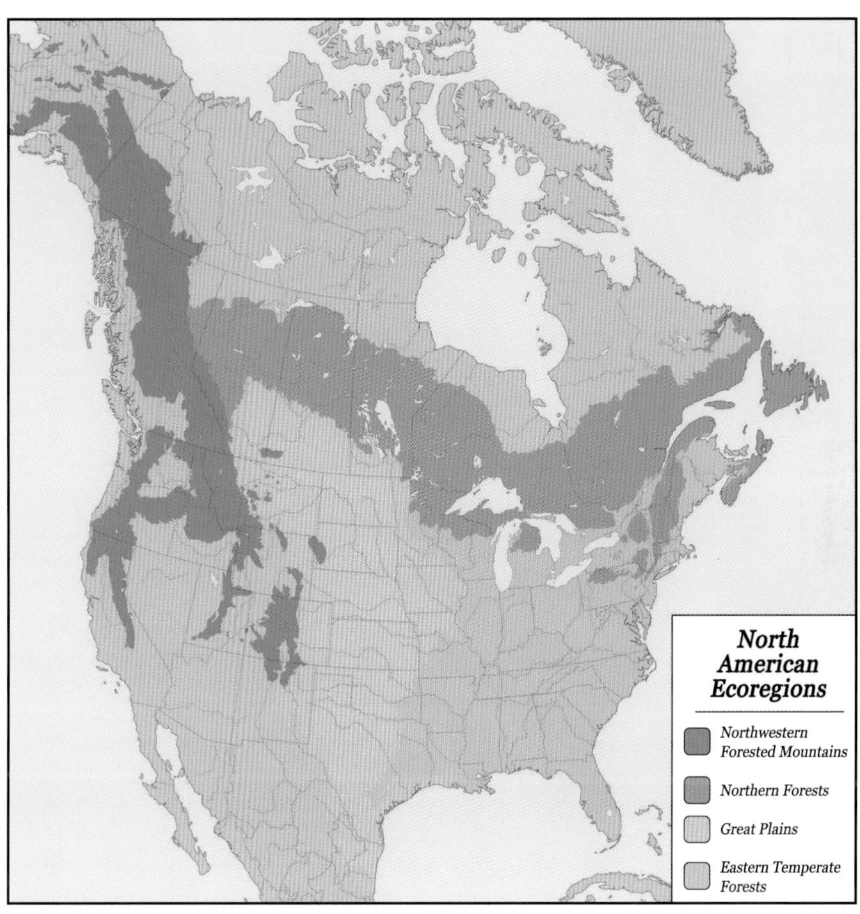


**Figure 33:** Distribution of North American ecoregions with relevance to the Salish migration accounts.

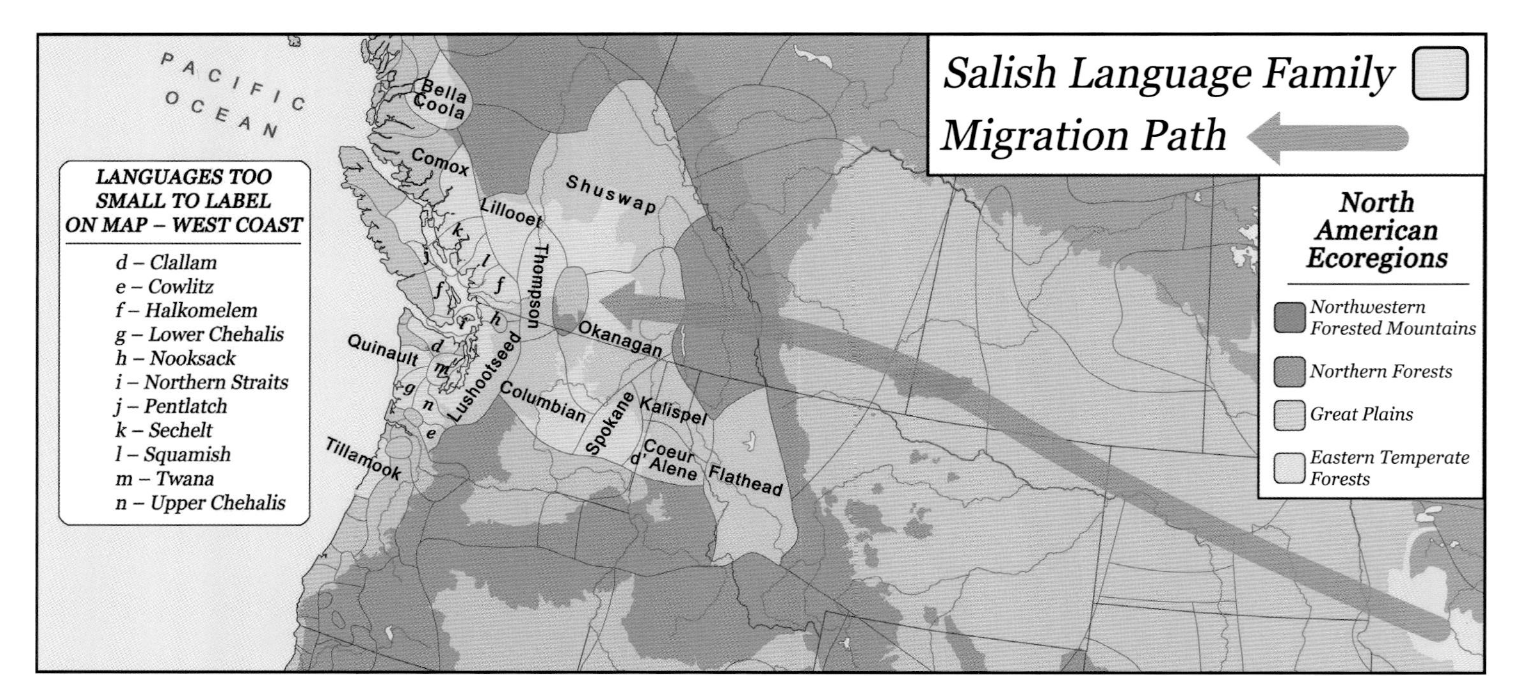


**Figure 34:** Possible migration path of Salish ancestors, superimposed on Contact-era geographic distribution of members of the Salish language family. North American ecoregions are also displayed. Where the ecoregions overlap the Contact-era distribution of peoples, the color displayed has been changed to a light green. Note that this green is distinct from the green used to display the "Eastern Temperate Forests" ecoregion.

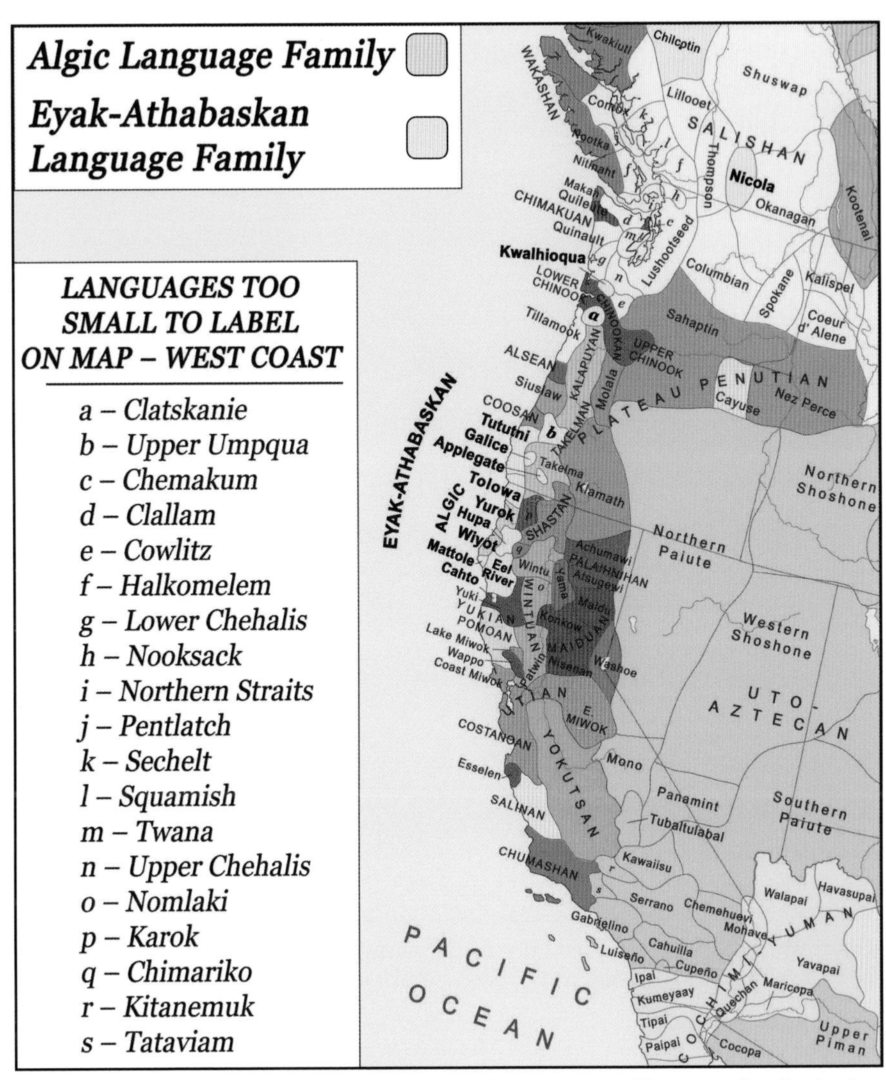


**Figure 35**: Contact-era geographic distribution of languages and language families in the western part of North America. The coasts of what are now Washington state, Oregon, and California show an especially Balkanized distribution of language families.

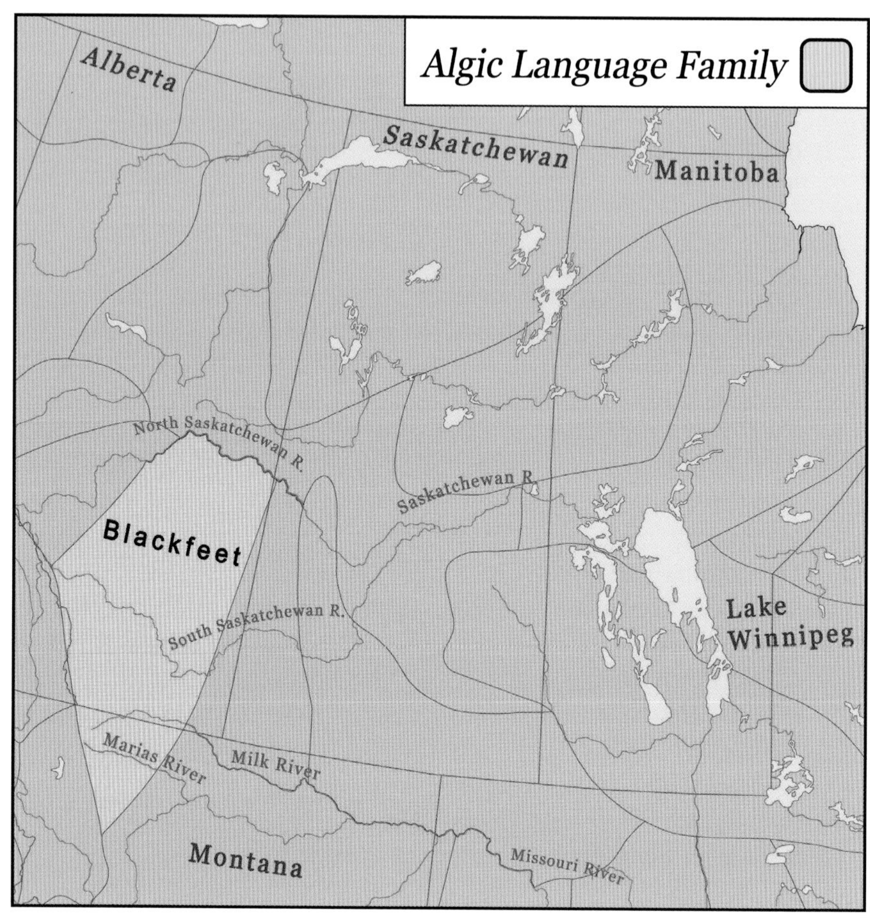


**Figure 36:** Contact-era geographic distribution of the Blackfeet. Also shown are geographic markers with relevance to ancestral Blackfeet migrations.

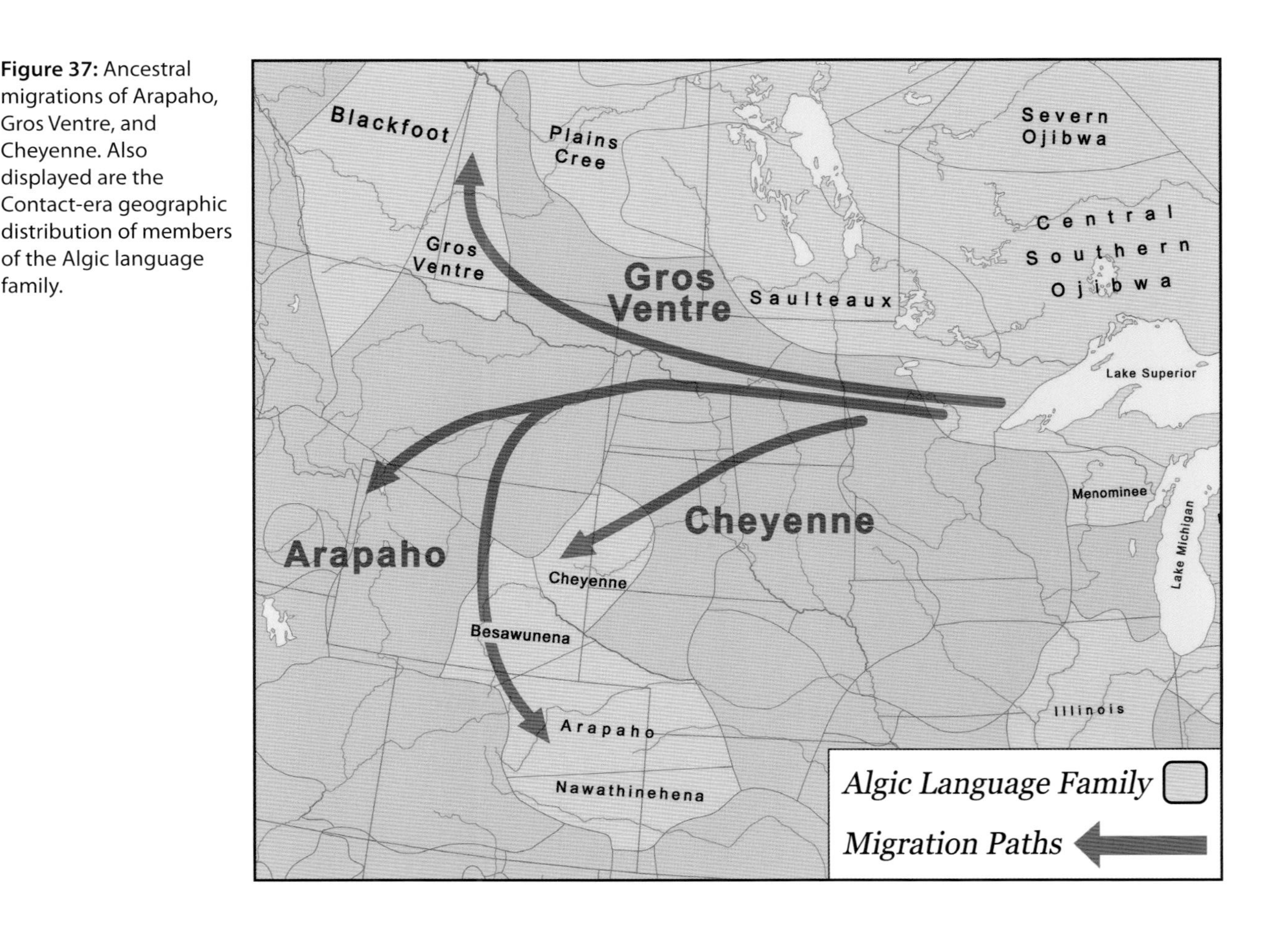


**Figure 37:** Ancestral migrations of Arapaho, Gros Ventre, and Cheyenne. Also displayed are the Contact-era geographic distribution of members of the Algic language family.

**Figure 38:** Battle sites for encounters between Algics and likely Athabaskans.

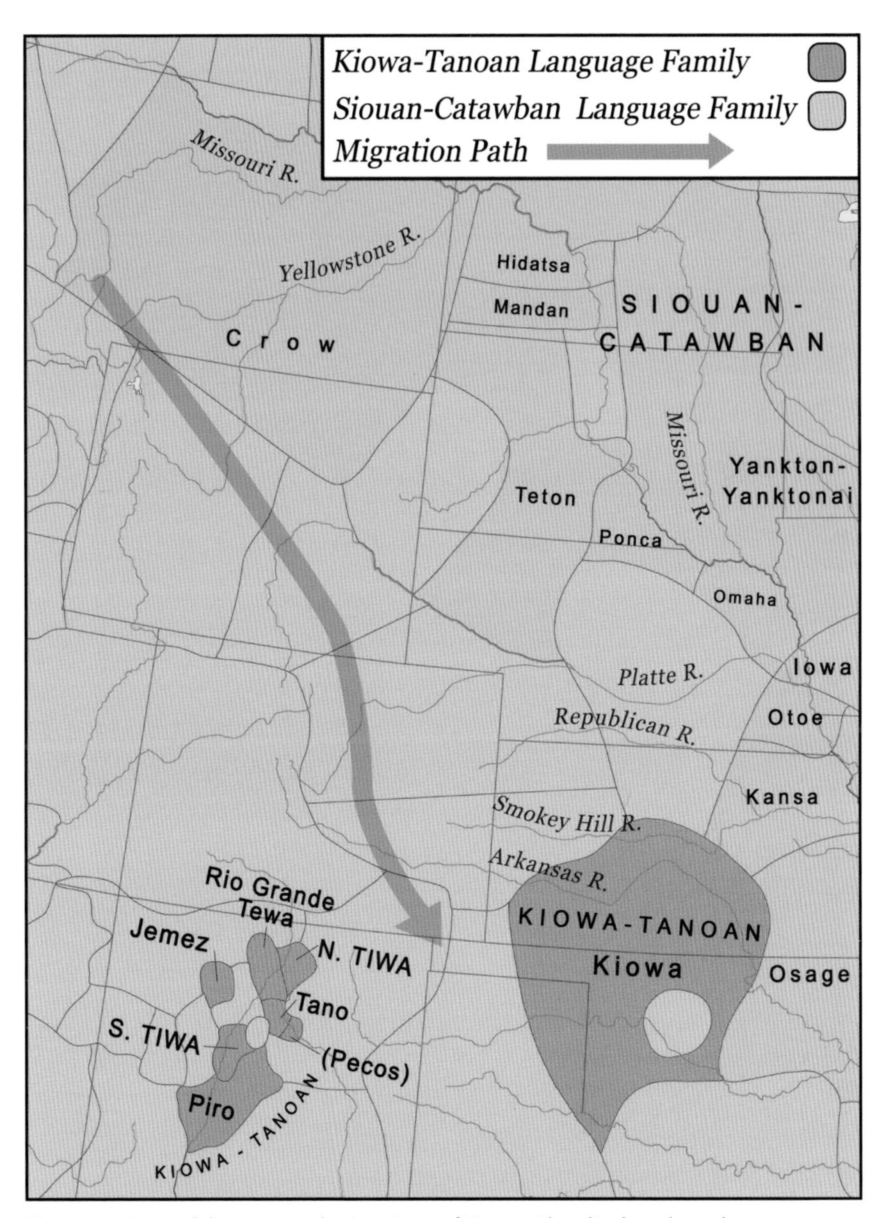


**Figure 39:** Map of the ancestral migrations of Kiowa. Also displayed are the Contact-era geographic distribution of members of the Kiowa-Tanoan and Siouan-Catawban language families.

**Figure 40:** Monk's Mound, the largest earthwork in the Americas. Located at Cahokia, just east of modern St. Louis.

**Figure 41:** Flat-topped mound in Aztalan State Park, Jefferson, Wisconsin.

**Figure 42:** Flat-topped mound in Moundville Archaeological Park, Moundville, Alabama.

**Figure 43:** Flat-topped mounds in Etowah Indian Mounds State Historic Site, Cartersville, Georgia.

**Figure 44:** Mayan temple at Chichen Itza, Mexico.

**Figure 45:** Reconstruction of Aztec Templo Mayor at Tenochtitlan (in what is now Mexico).

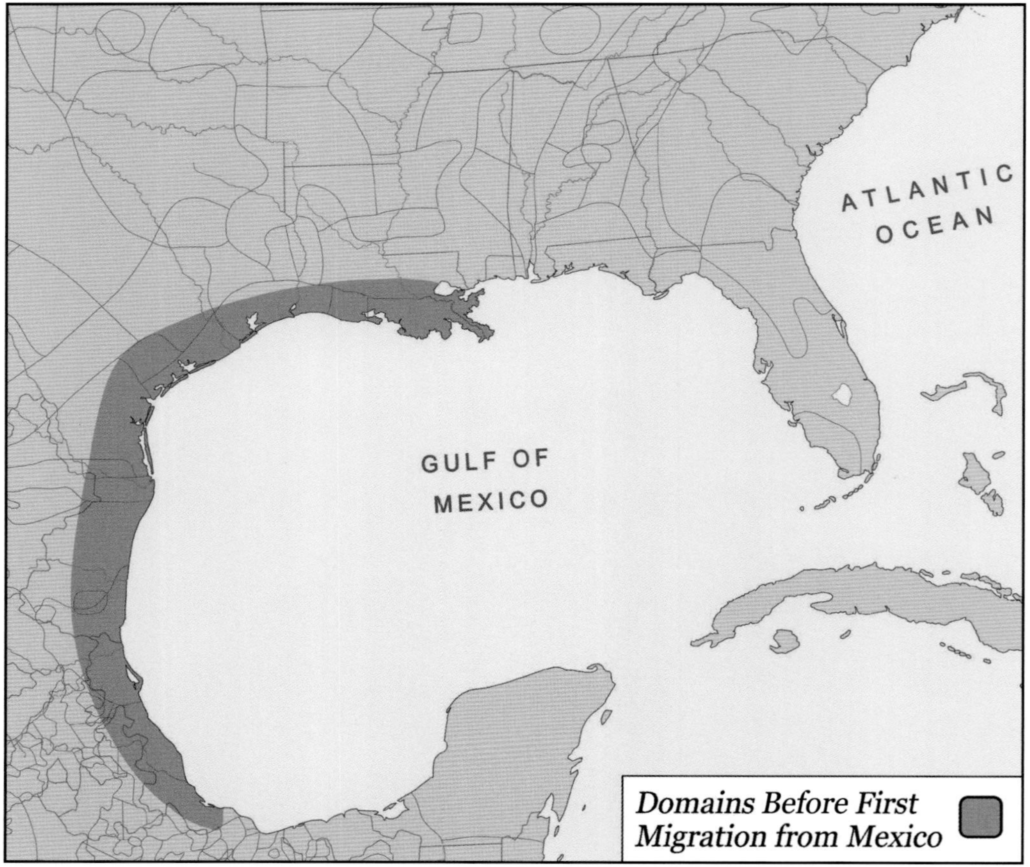

**Figure 46:** Former domains of Contact-era Native tribe: early phase.

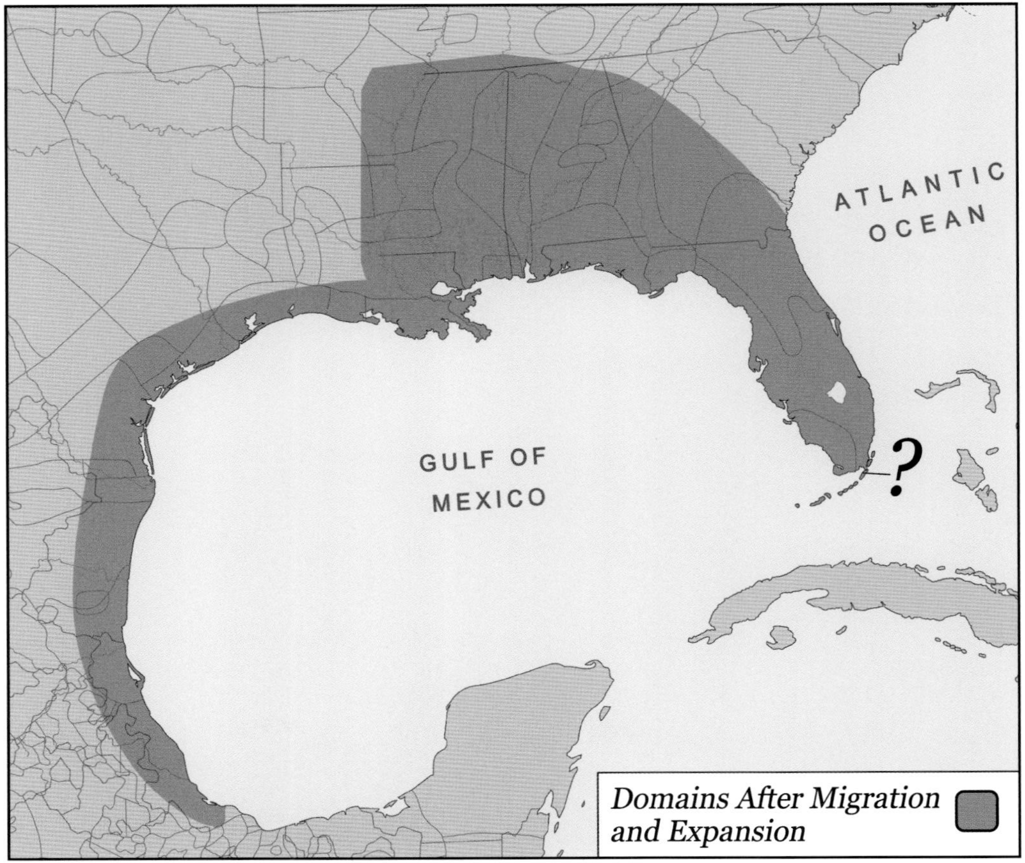

**Figure 47:** Former domains of Contact-era Native tribe: early migration and expansion.

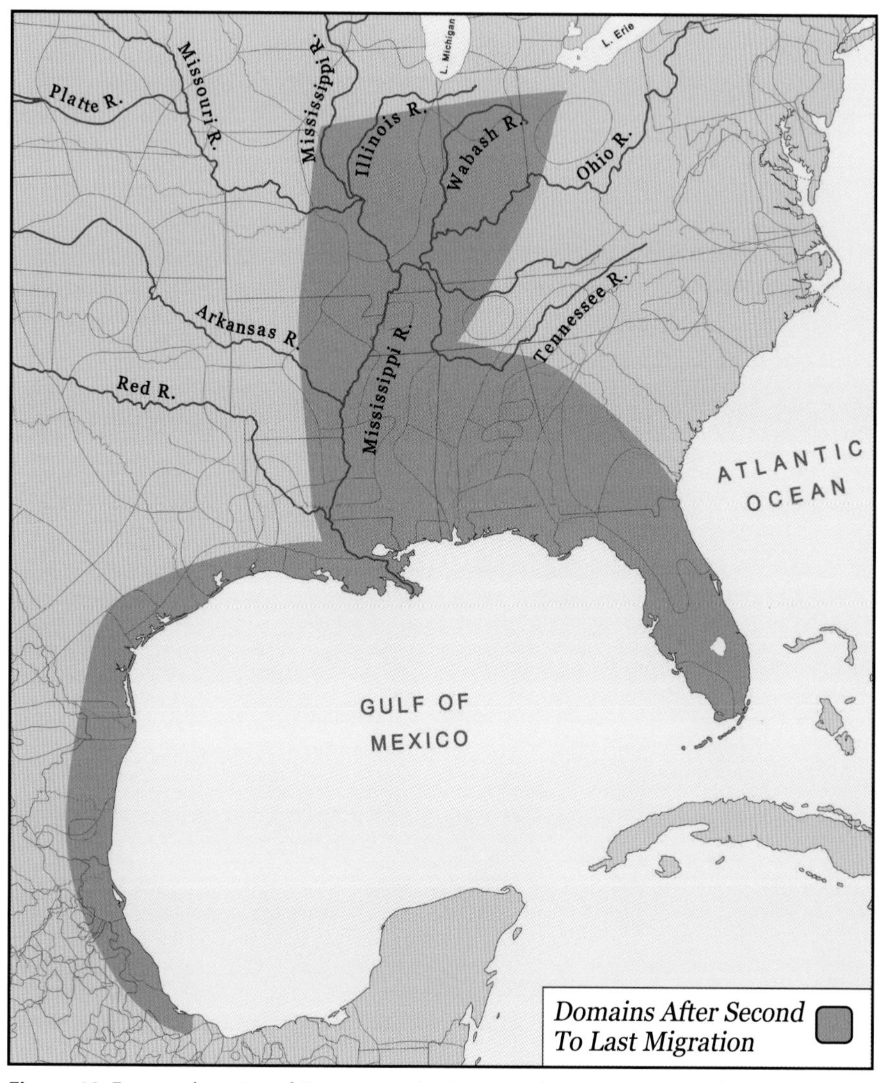


**Figure 48:** Former domains of Contact-era Native tribe: later migration and expansion.

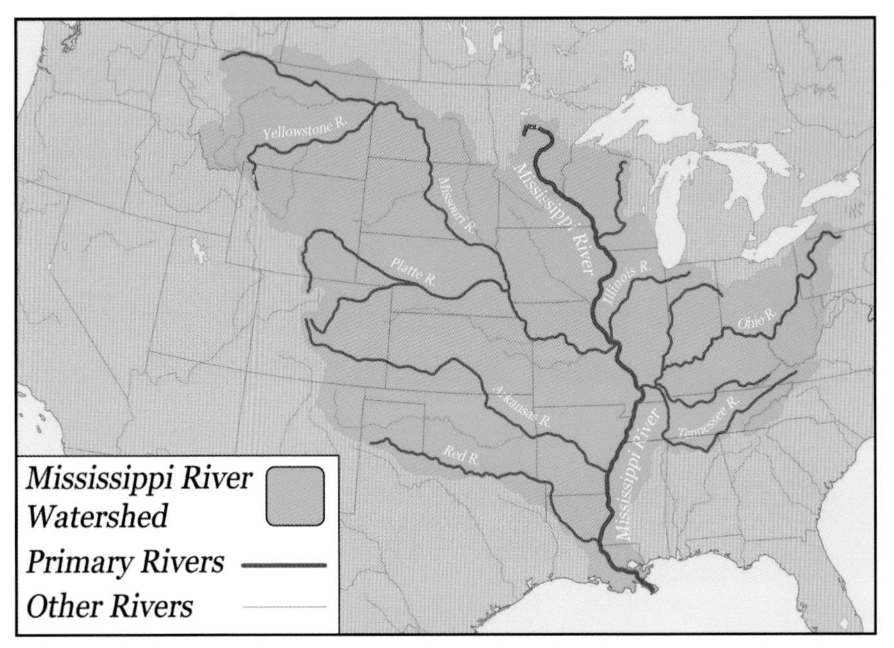


**Figure 49:** Watershed of the Mississippi River.

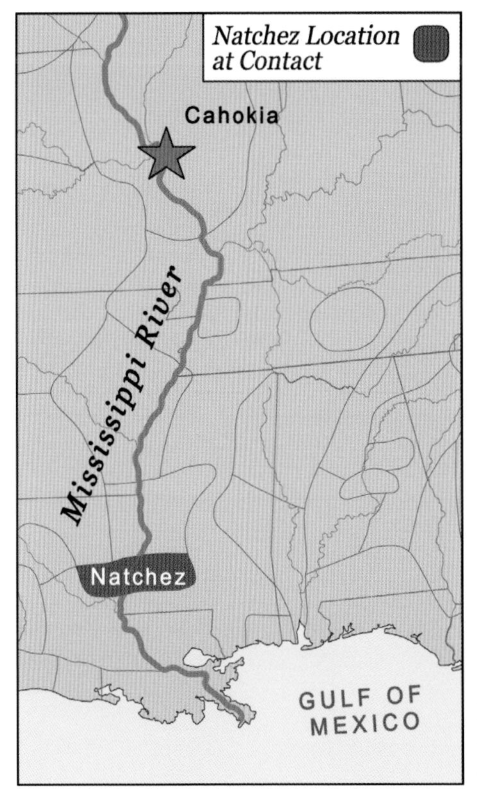


**Figure 50:** Contact-era geographic distribution of the Natchez.

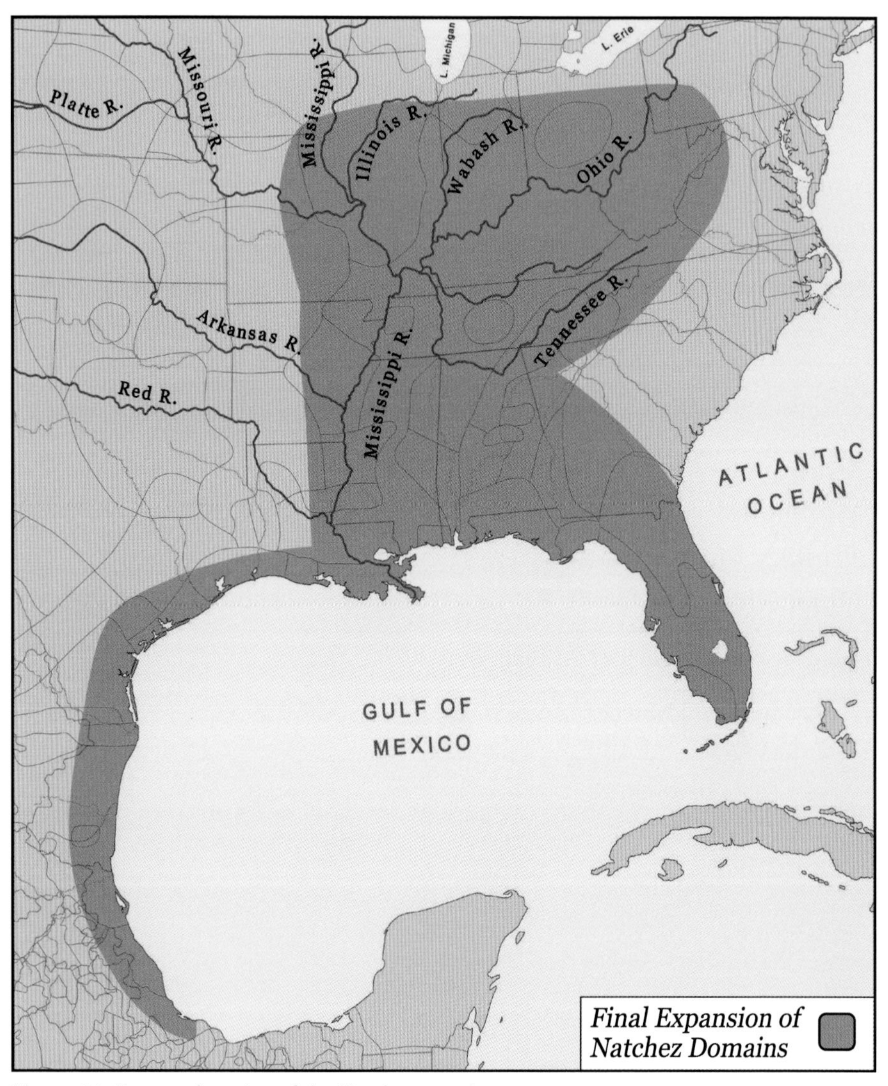


**Figure 51:** Former domains of the Natchez: maximum extent.

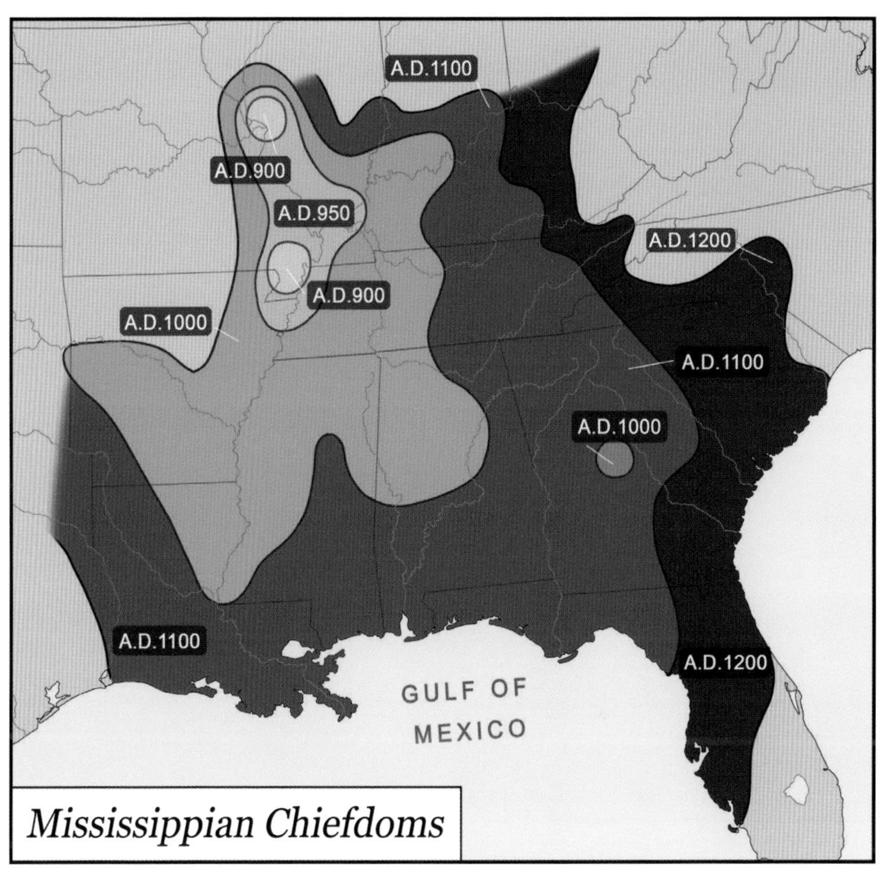


**Figure 52:** Pre-Contact growth and expansion of Native chiefdoms.

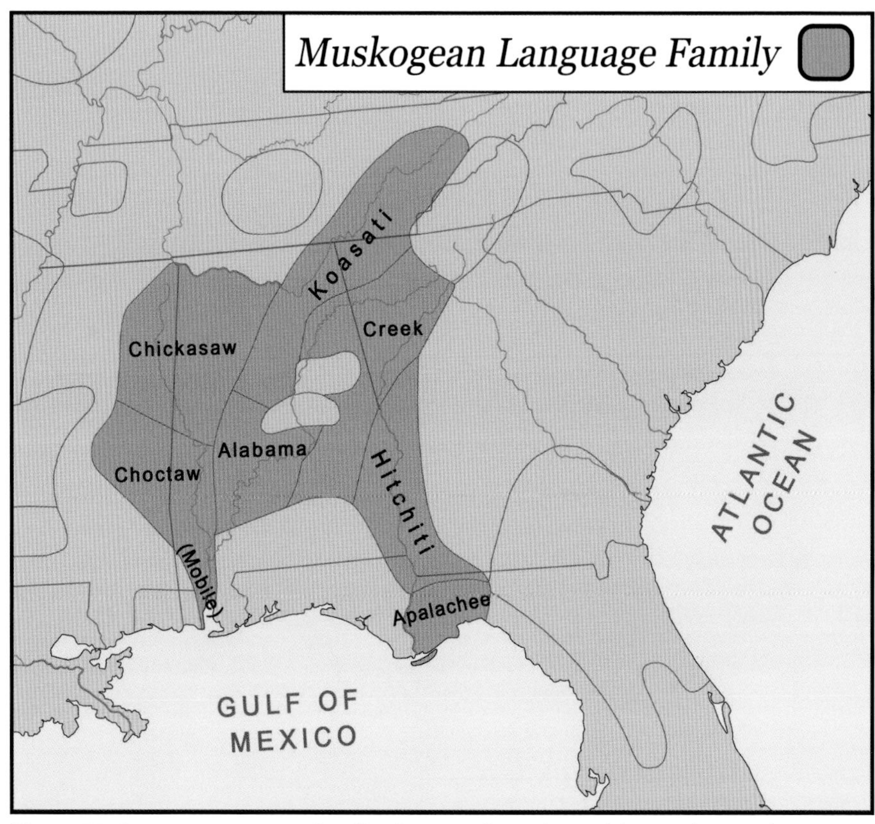


**Figure 53:** Contact-era geographic distribution of members of the Muskogean language family.

**Figure 54:** Partial map of De Soto's A.D. 1500s-era migration through the Southeast. Note that De Soto's men encountered likely Muskogeans—"Apalachee"–in what is now northern Florida and "Chicaza" (Chicksaw?) in what is now eastern Mississippi.

**Figure 55:** Eastern face of Monk's Mound, located in Cahokia, just east of modern St. Louis.

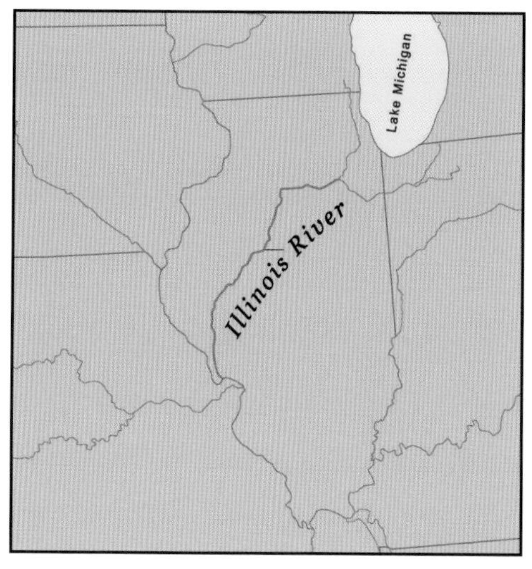


**Figure 56:** Map of Illinois River, which flows into the Mississippi just north of modern St. Louis.

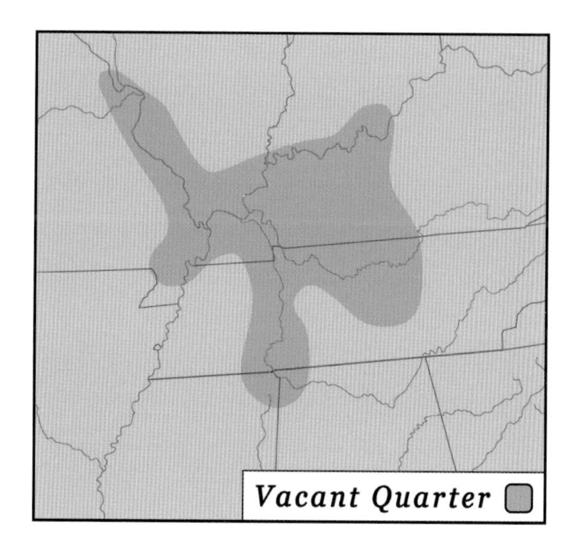


**Figure 57:** Map of the *Vacant Quarter*, which extends from Cahokia south and east.

**Figure 58:** Contact-era geographic distribution of members of the Siouan-Catawban language family.

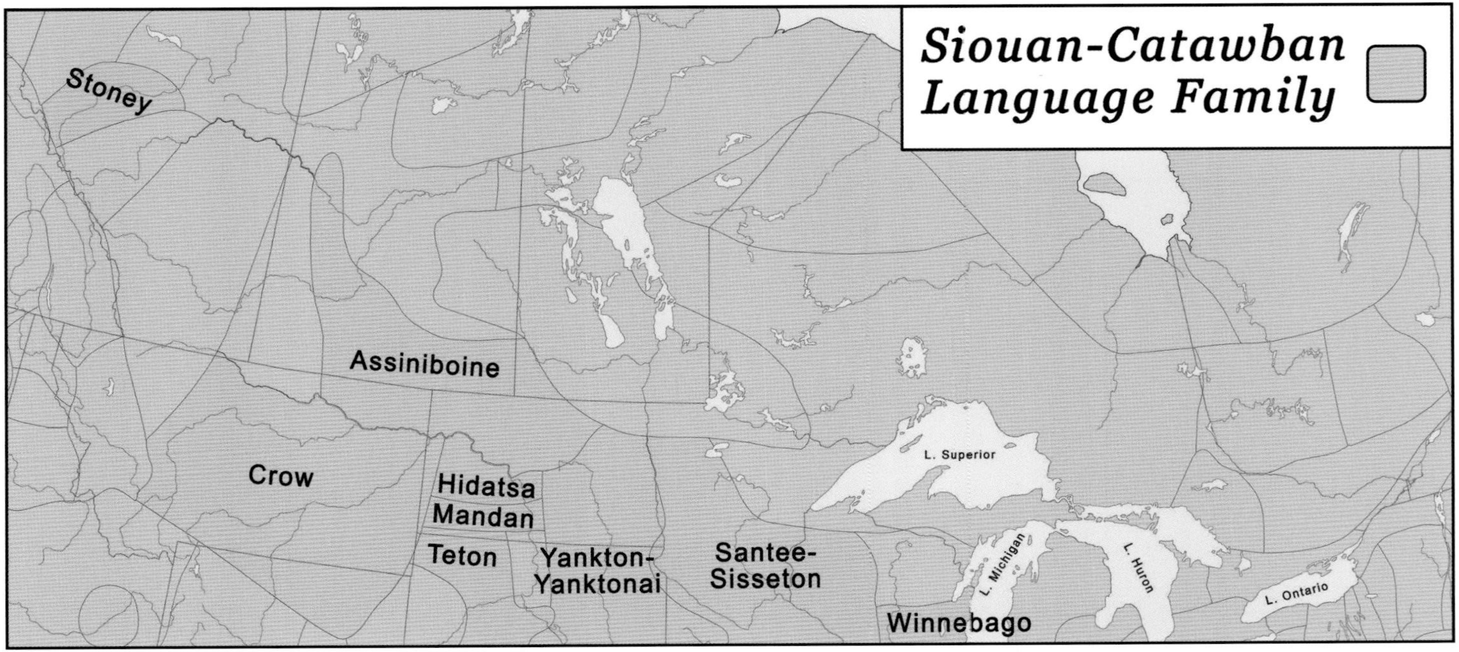

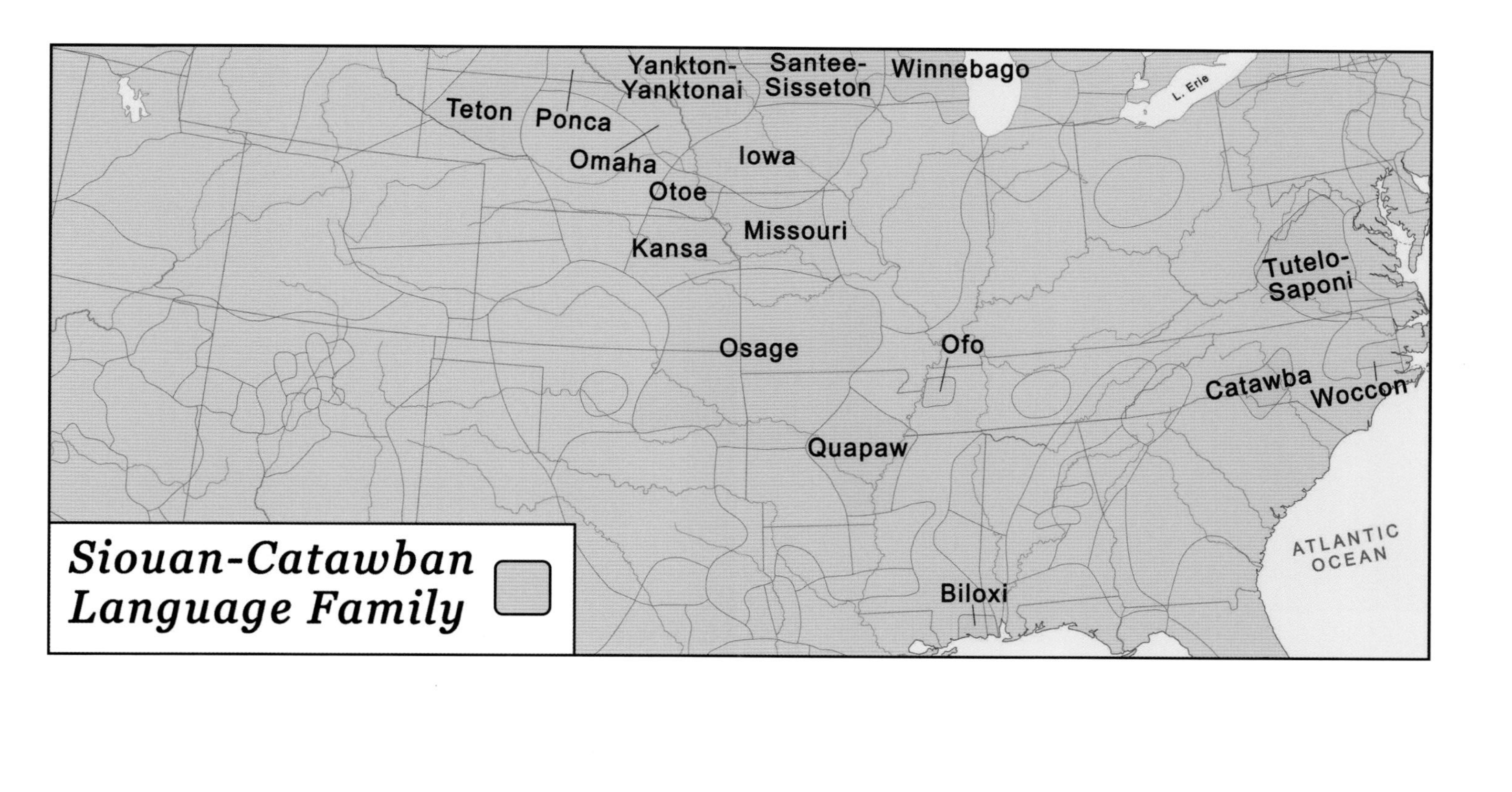
Yankton-
Yanktonai
Santee-
Sisseton
Winnebago
L. Erie
Teton
Ponca
Omaha
Iowa
Otoe
Missouri
Kansa
Tutelo-
Saponi
Osage
Ofo
Catawba
Woccon
Quapaw
ATLANTIC
OCEAN
Biloxi
Siouan-Catawban
Language Family

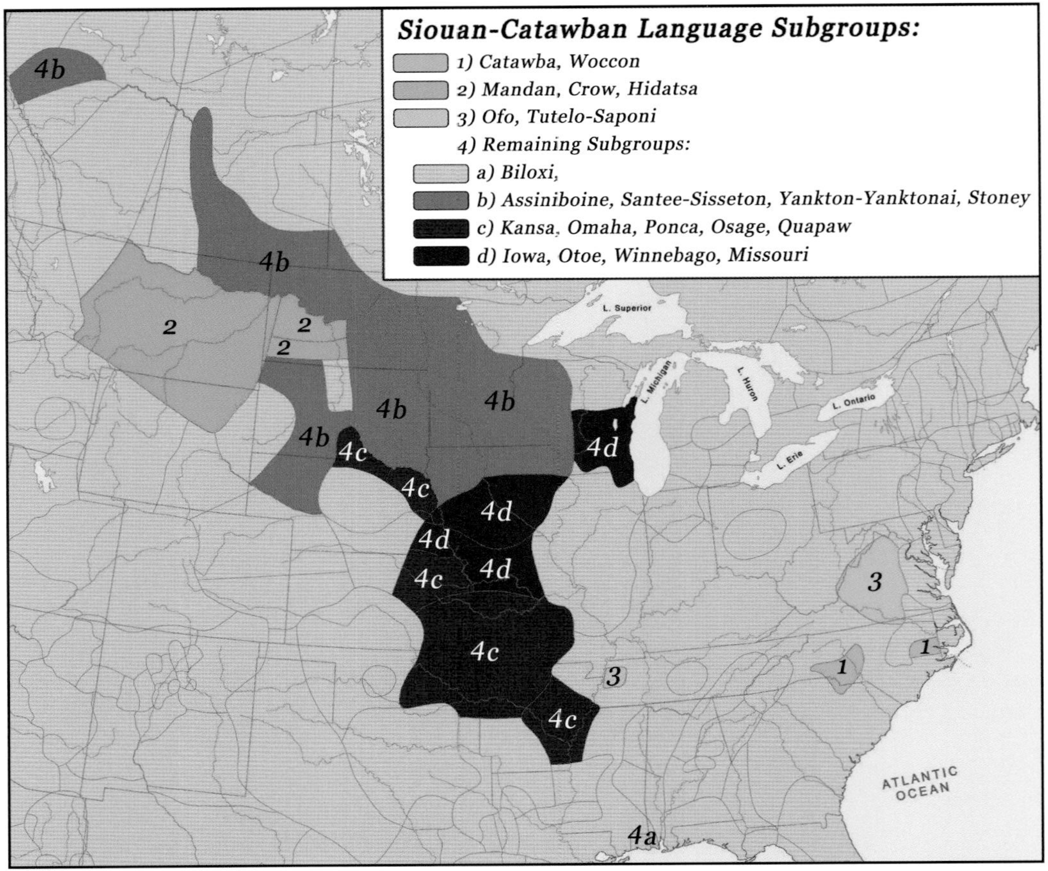

**Figure 59:** Contact-era geographic distribution of members of the Siouan-Catawban language family, now colored according to linguistic subgroup. The linguistic subgroups are numbered from oldest (1) to youngest (4).

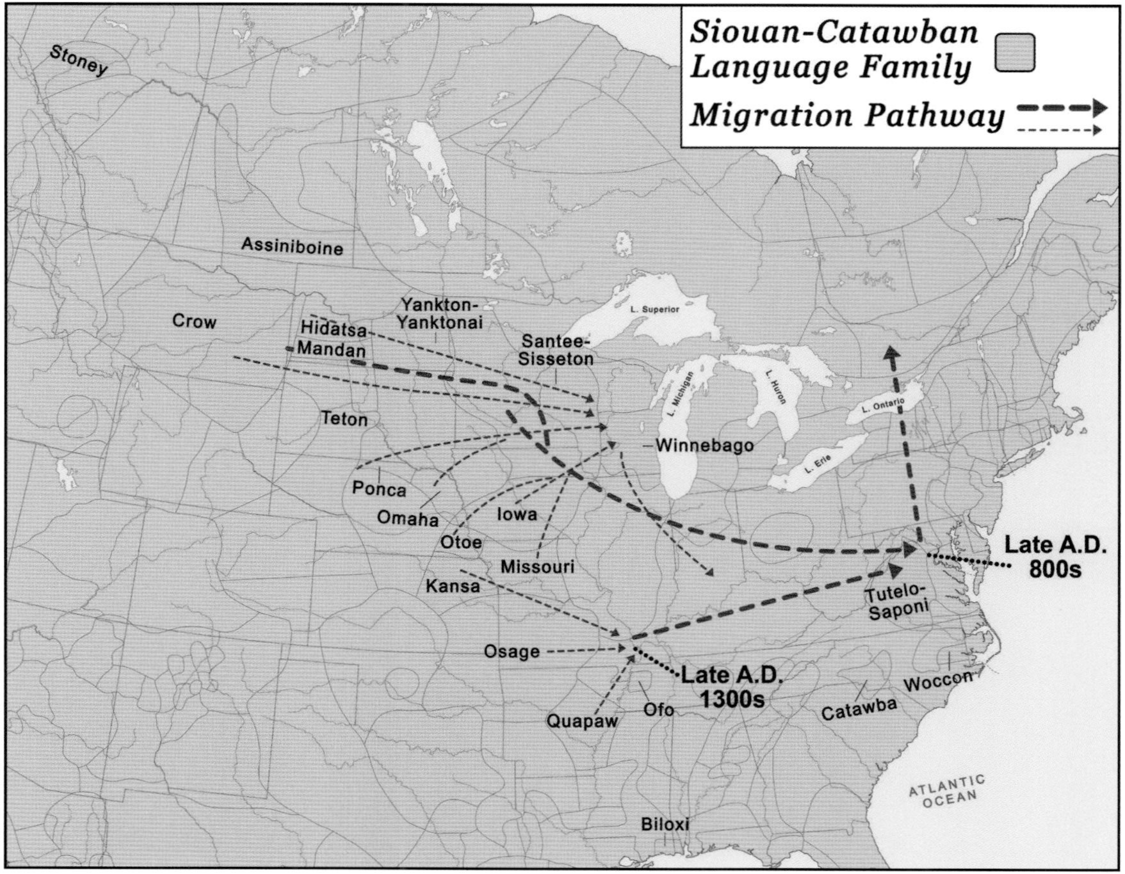

**Figure 60:** Derivation of the ancestral homelands of members of the Siouan-Catawban language family, as per their own histories. The migration pathways crisscross each other frequently. Therefore, both thick and thin dashed lines are used to make it easier to follow any individual pathway.

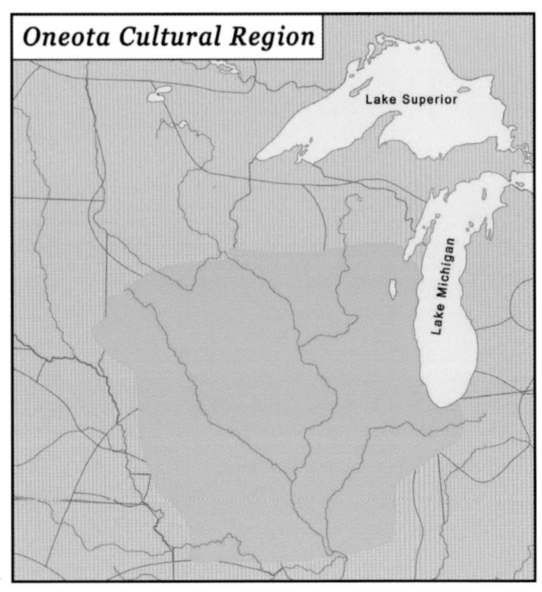


**Figure 61:** Map of Oneota culture area.

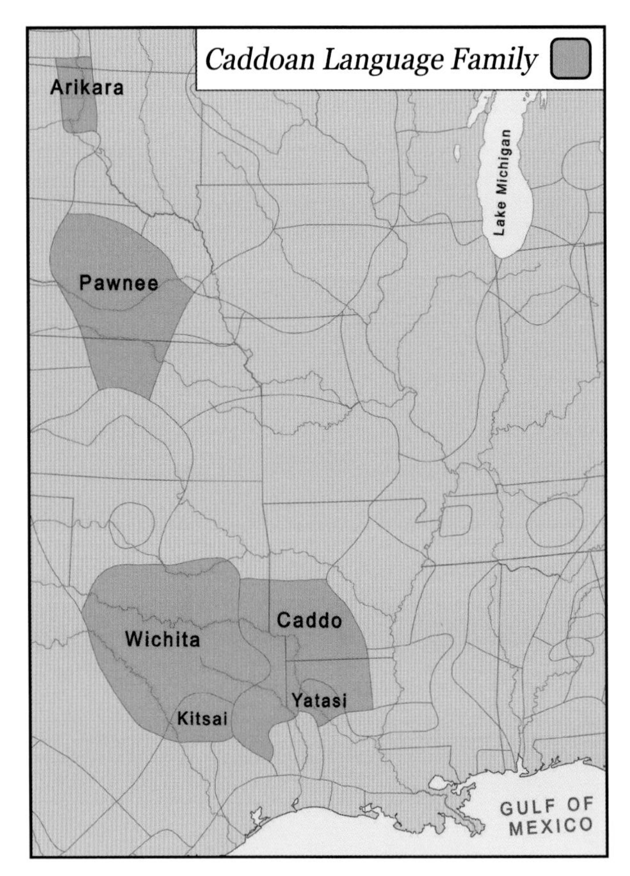


**Figure 62:** Contact-era geographic distribution of members of the Caddoan language family.

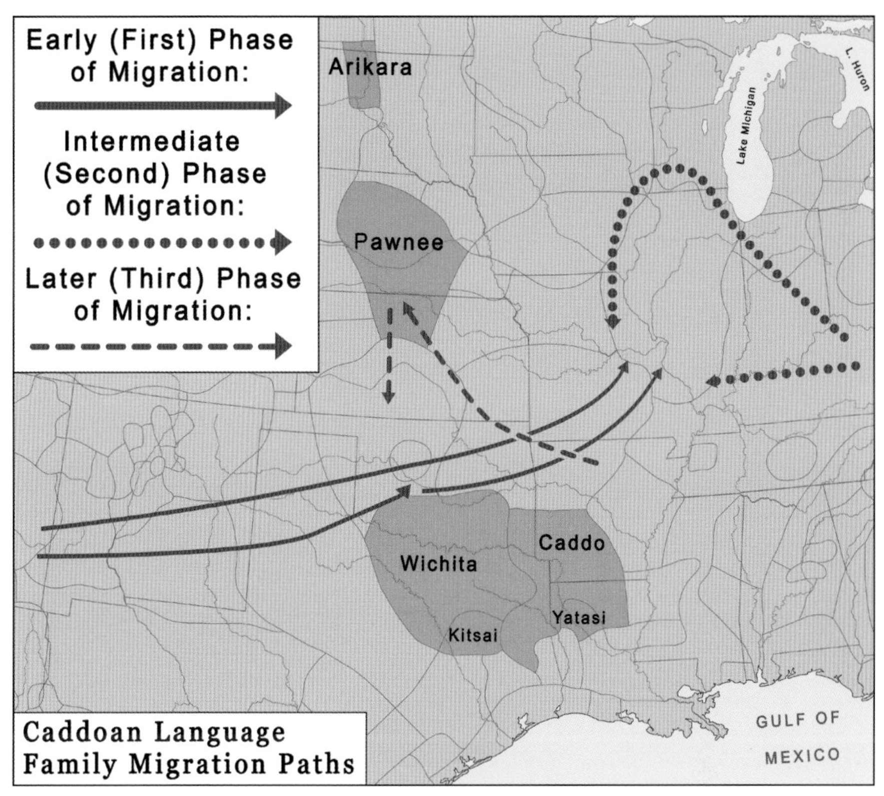


**Figure 63:** Ancestral migrations of members of the Caddoan language family, as per their own histories. Also shown is their Contact-era geographic distribution.

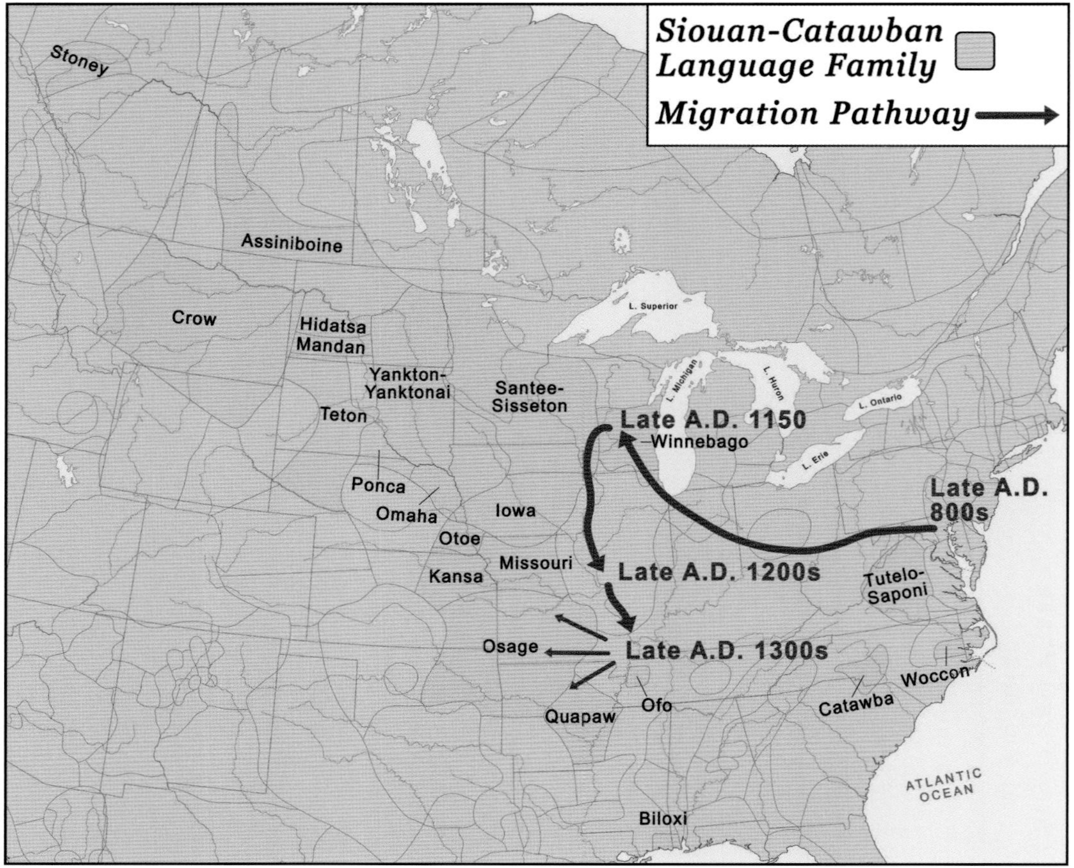

**Figure 64:** Reconstruction of the ancestral Siouan-Catawban migrations.

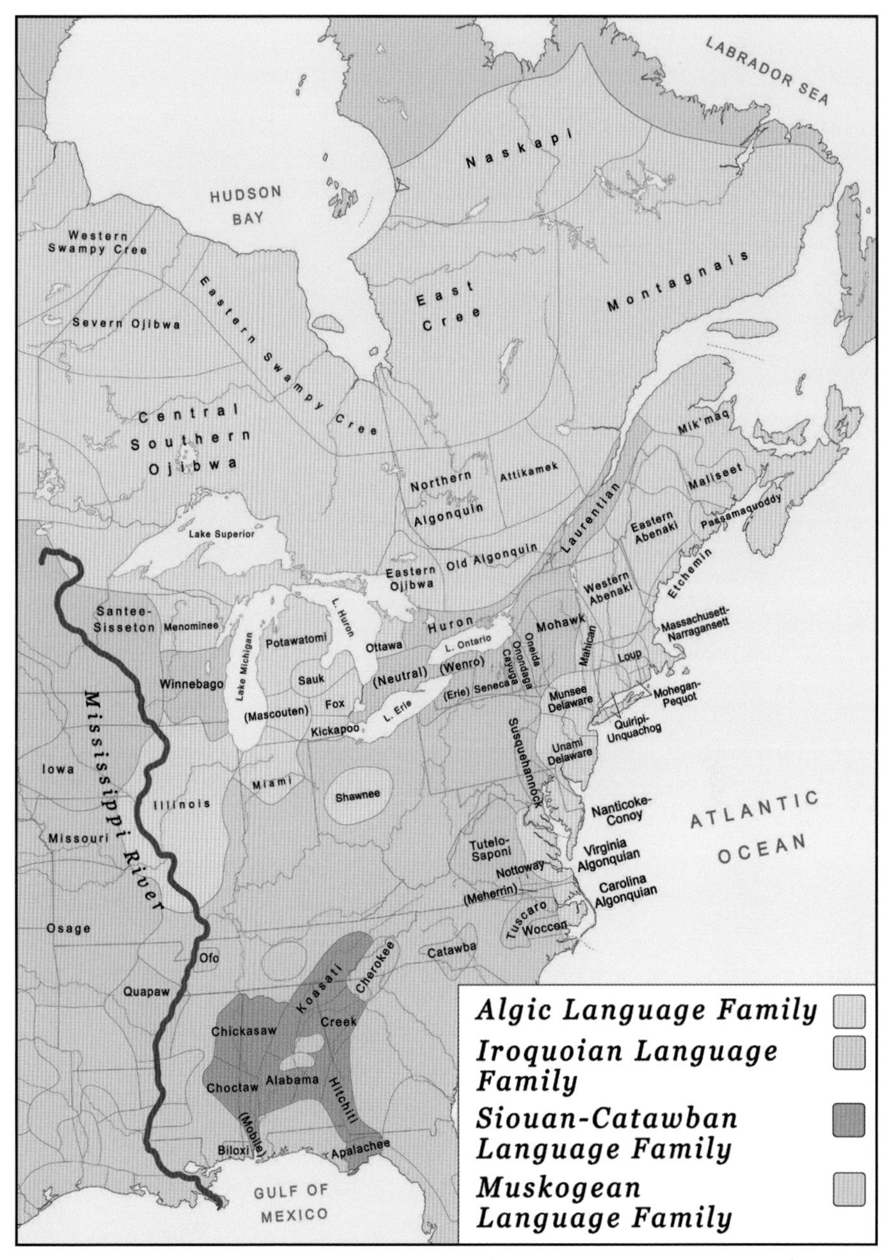


**Figure 65:** Contact-era geographic distribution of major language groups east of the Mississippi River.

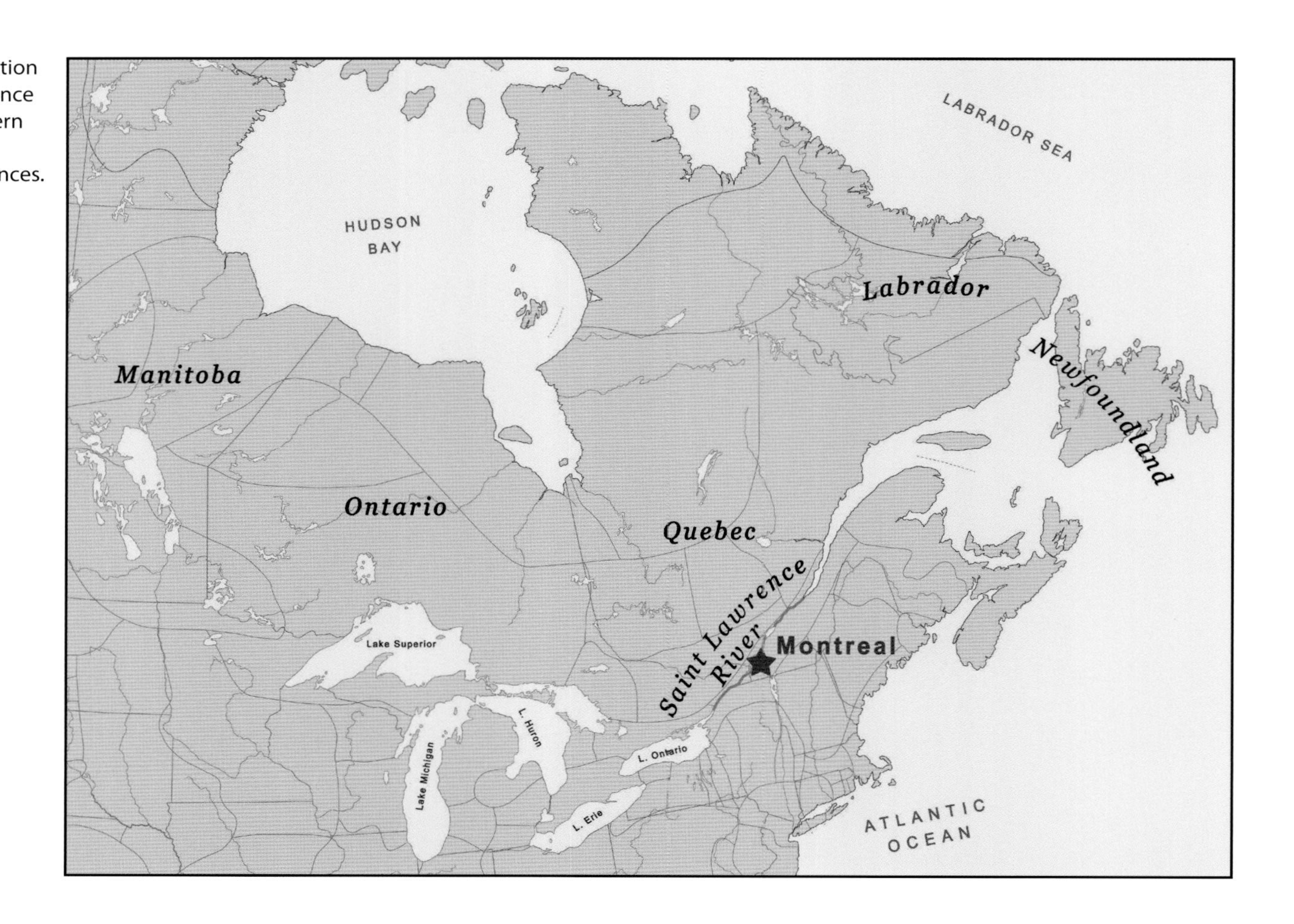


**Figure 66:** Location of the St. Lawrence River and modern Montreal and Canadian provinces.

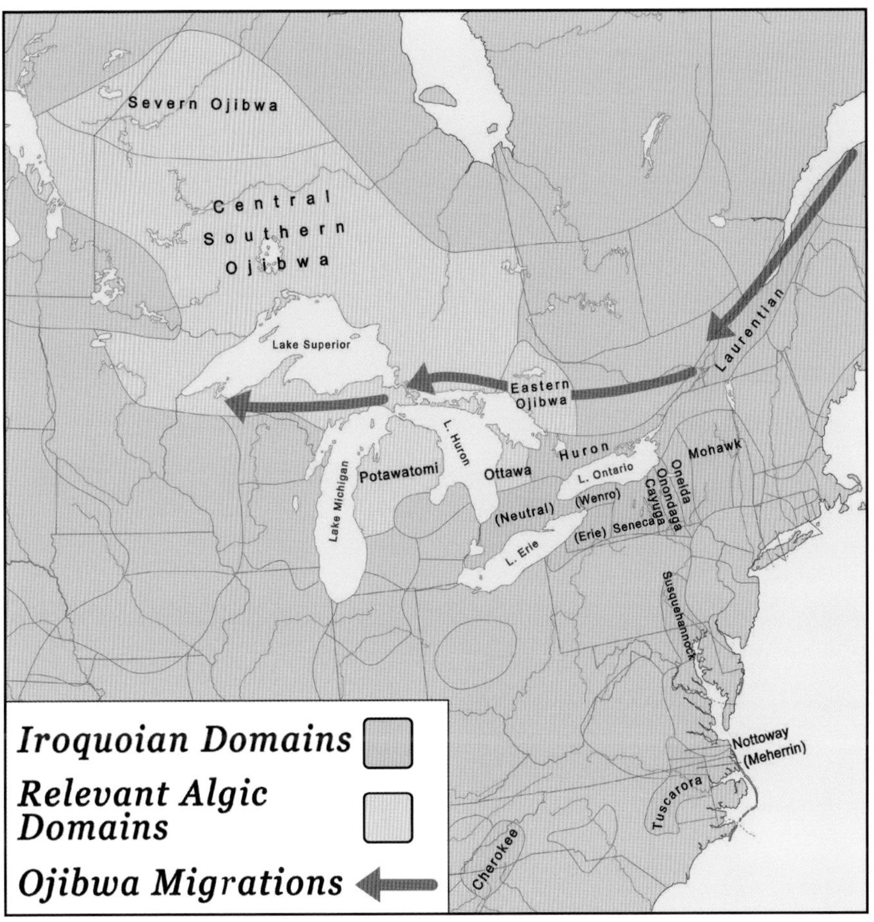


**Figure 67:** Map of the Great Lakes region with Ojibwa ancestral migrations displayed. The first arrow on the left ends near where modern Montreal sits. The last arrow on the right begins near Michilimackinac and ends near La Pointe. Also shown are the Contact-era locations of some of the members of the Algic and Iroquoian language families.

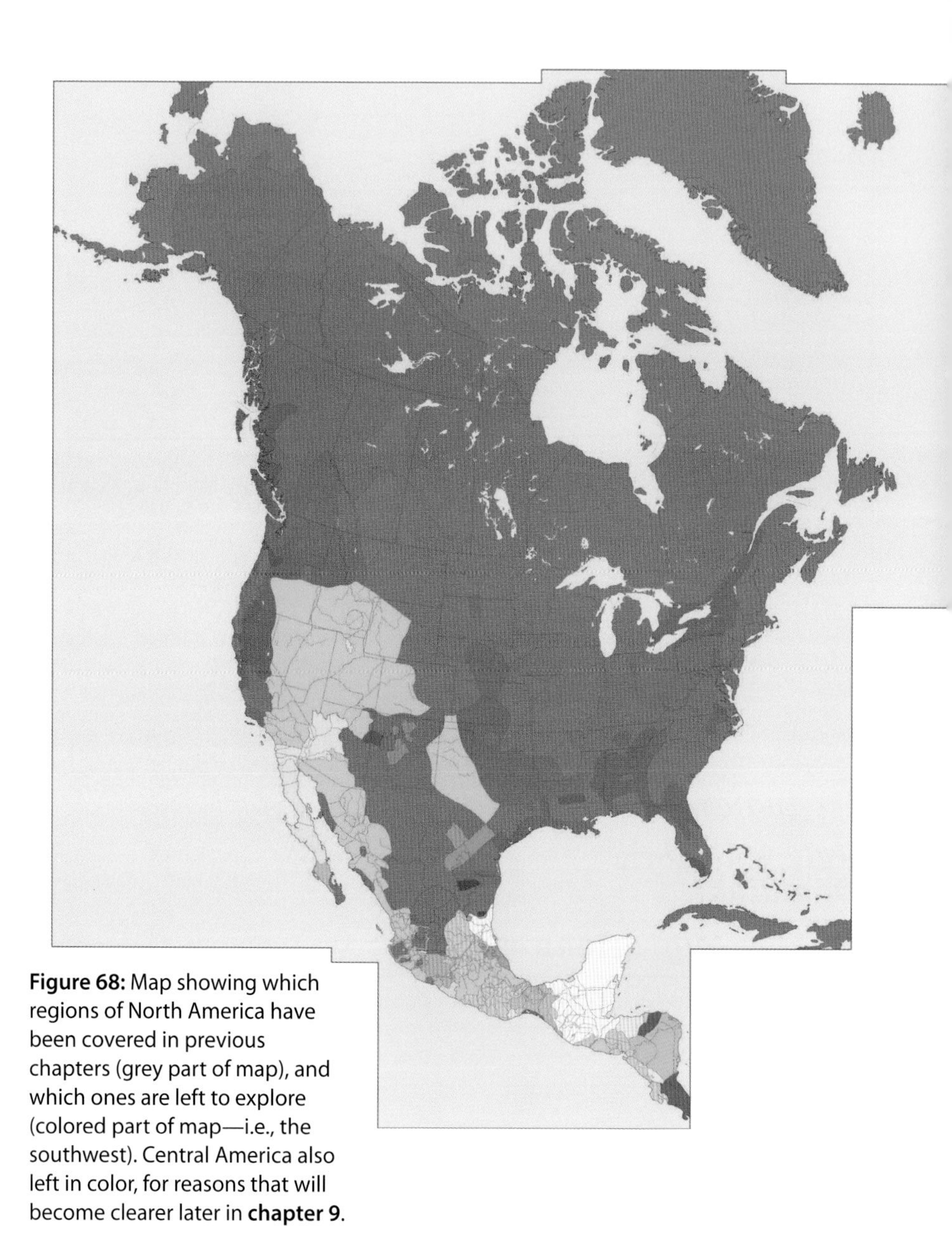

**Figure 68:** Map showing which regions of North America have been covered in previous chapters (grey part of map), and which ones are left to explore (colored part of map—i.e., the southwest). Central America also left in color, for reasons that will become clearer later in **chapter 9**.

**Figure 69:** Walnut Canyon, just east of Flagstaff, Arizona.

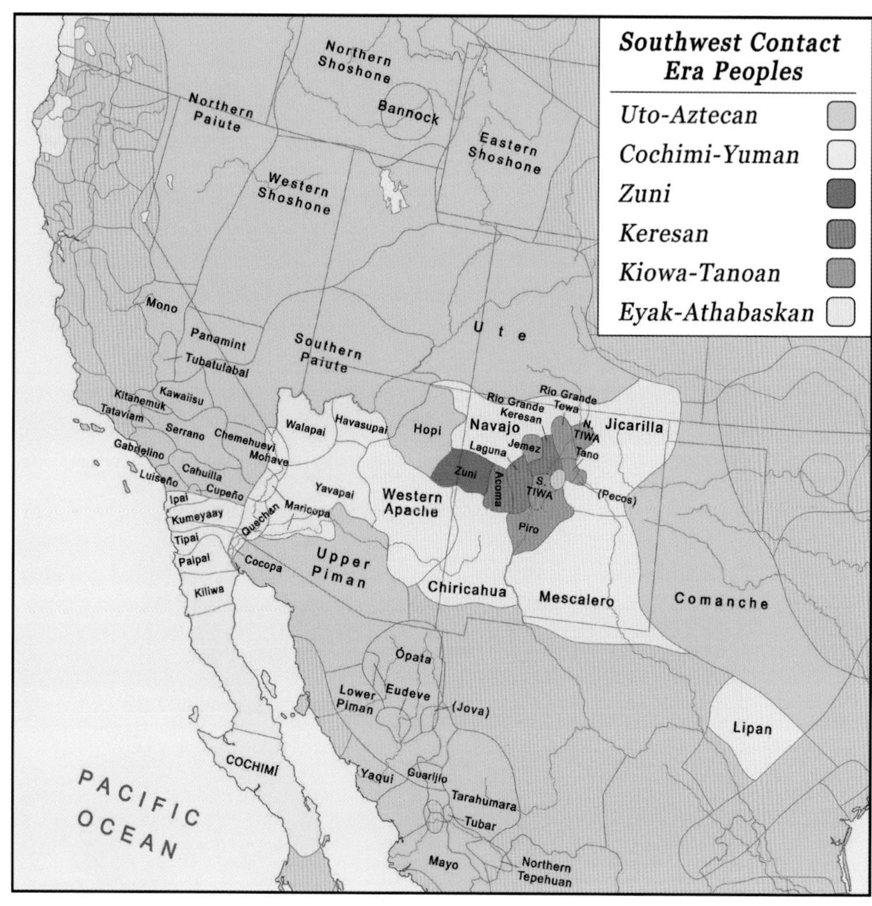


**Figure 70:** Contact-era geographic distribution of members of various language families in the Southwest.

**Figure 71:** Ruins at Montezuma Castle, south of Flagstaff, Arizona.

**Figure 72:** Ruins at Tuzigoot National Monument, south of Flagstaff, Arizona.

**Figure 73:** Ruins at Wupatki National Monument, north of Flagstaff, Arizona.

**Figure 74:** Ruins at Wukoki, part of Wupatki National Monument, north of Flagstaff, Arizona.

**Figure 75:** Ruins of the Citadel, at Wupatki National Monument, north of Flagstaff, Arizona.

**Figure 76:** Ruins of a house at the edge of Box Canyon, at Wupatki National Monument, north of Flagstaff, Arizona.

**Figure 77:** Cliff Palace, at Mesa Verde National Monument, in southwestern Colorado.

**Figure 78:** Foreground: *Kin Kletso* ruins. Background: Canyon, Chaco Wash, and canyon walls in Chaco Canyon National Historical Park. Location: Northwestern New Mexico.

**Figure 79:** Ruins of Pueblo Bonito, at Chaco Canyon National Historical Park, in northwestern New Mexico.

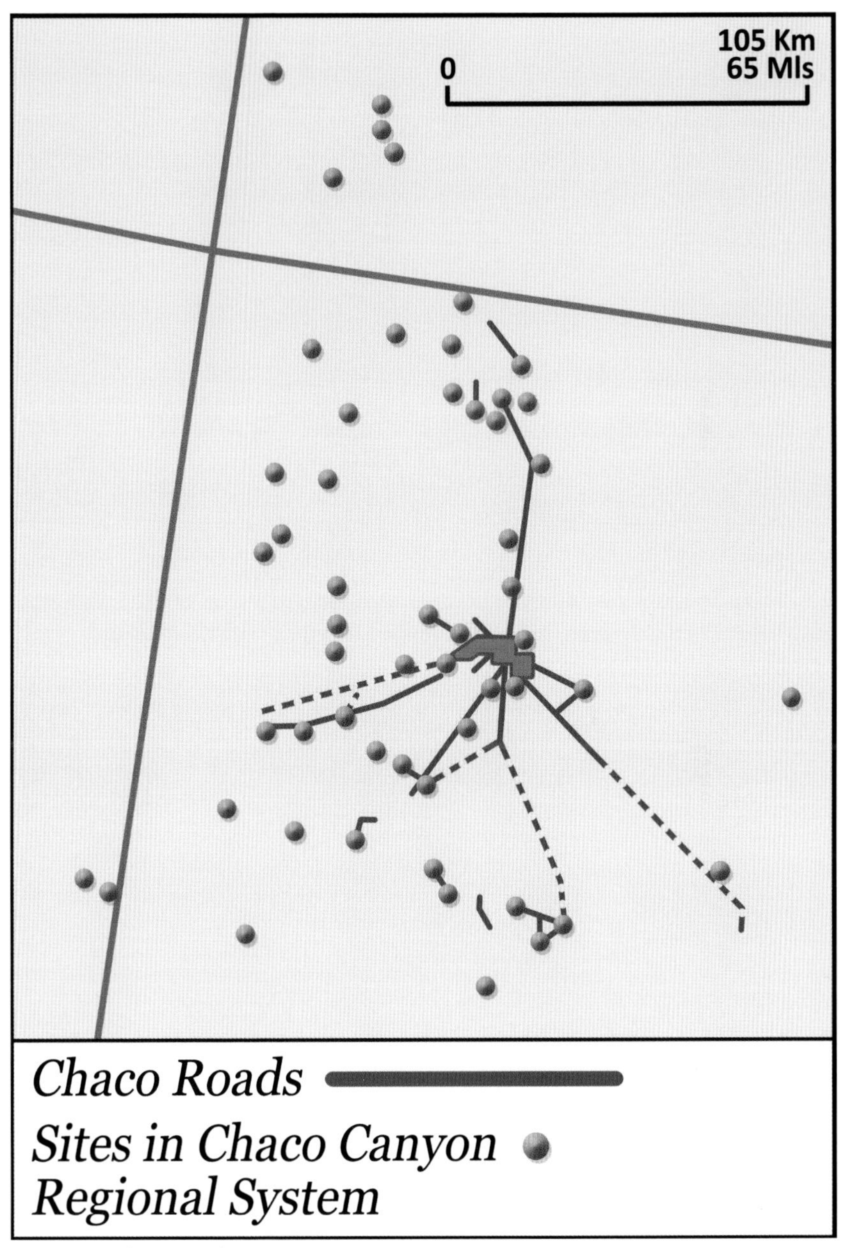


**Figure 80:** Chaco road system.

**Figure 81:** Ruins of the Great House at Aztec Ruins, in northwestern New Mexico.

**Figure 82:** Example of Chaco-style masonry at Aztec Ruins, in northwestern New Mexico.

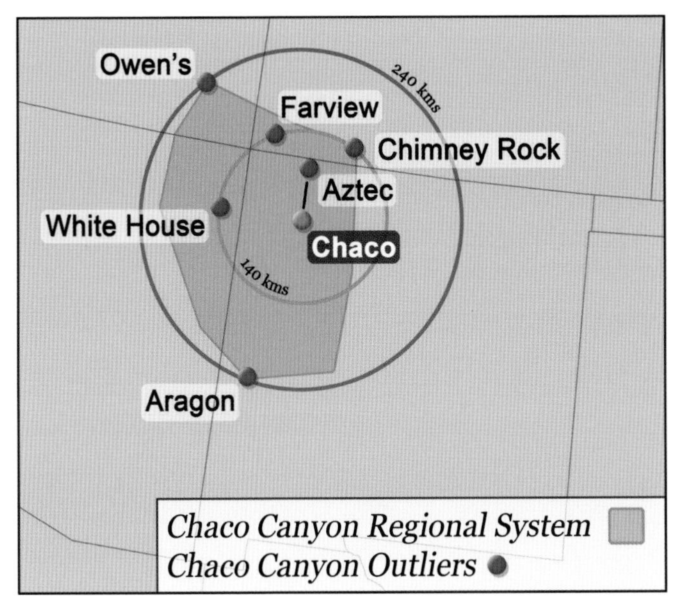


**Figure 83:** Map of Chaco culture network.

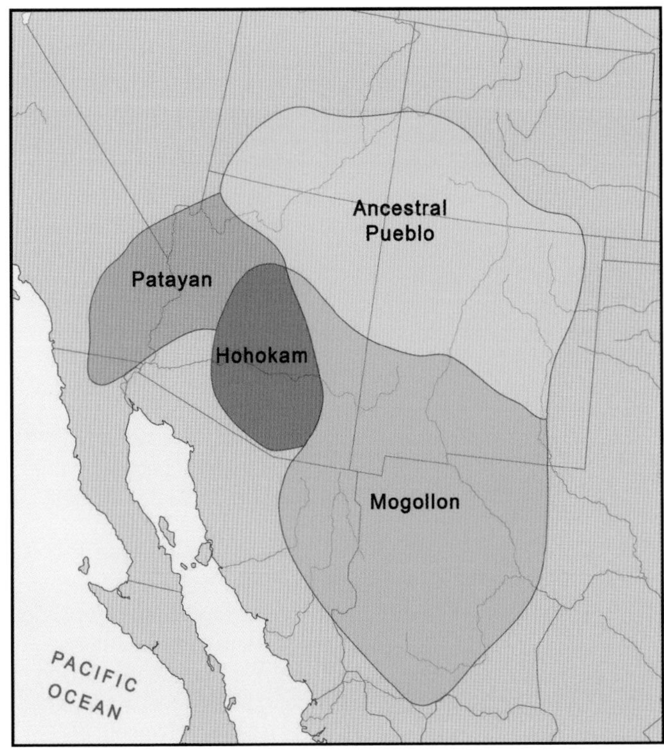


**Figure 84:** Major cultural areas in the pre-Contact southwest.

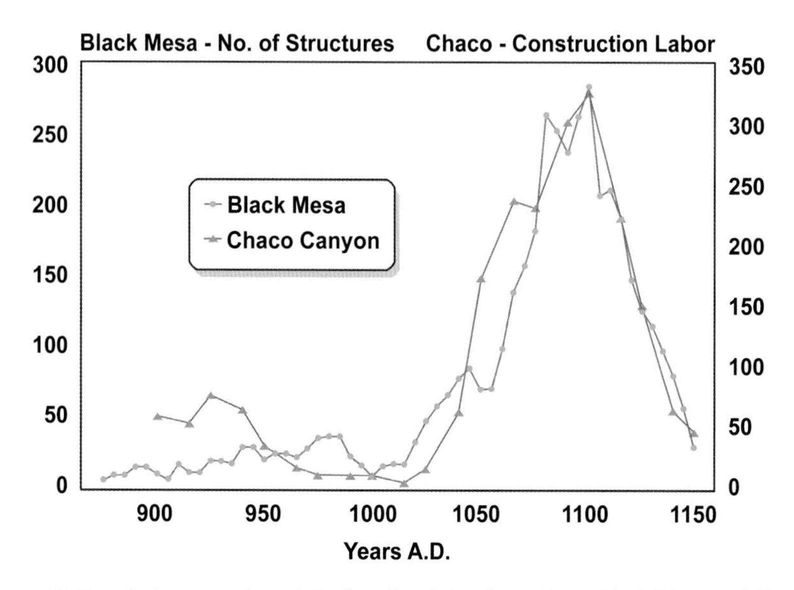


**Figure 85:** Population growth and decline: Parallel trajectories at Black Mesa and Chaco Canyon.

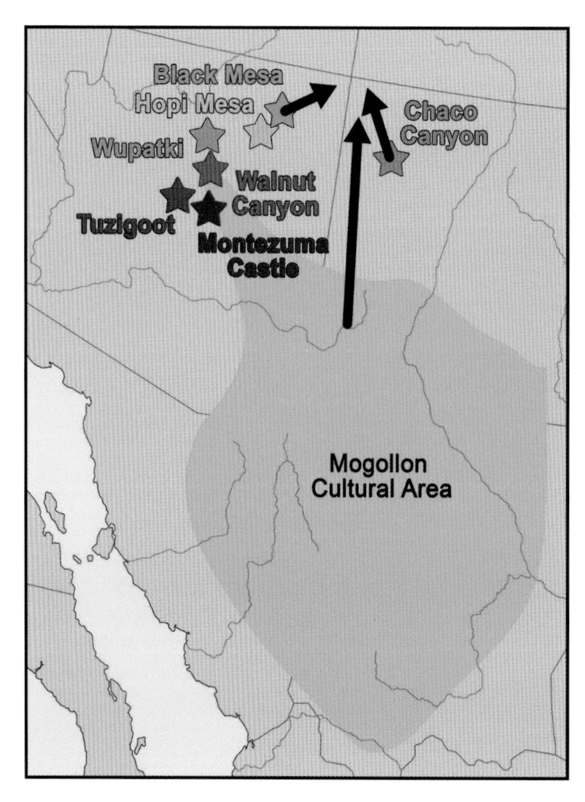


**Figure 86:** Directional movement of peoples in the late pre-Contact southwest. Black arrow on the left represents movement of peoples from Black Mesa; black arrow on the right represents movement of peoples from Chaco Canyon. Long black arrow in the center represents movement of peoples from the northern Mogollon cultural area.

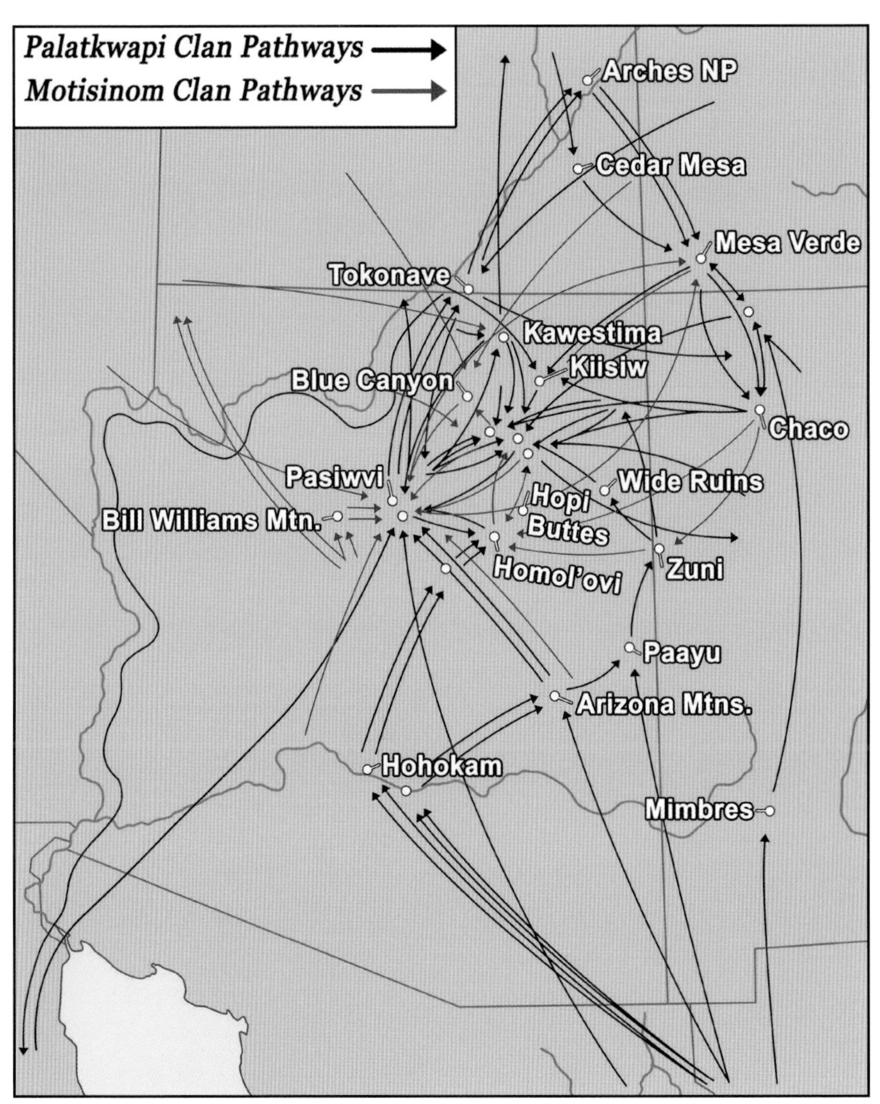


**Figure 87:** Map of ancestral Hopi migrations.

**Figure 88:** Ball court at Wupatki National Monument, north of Flagstaff, Arizona.

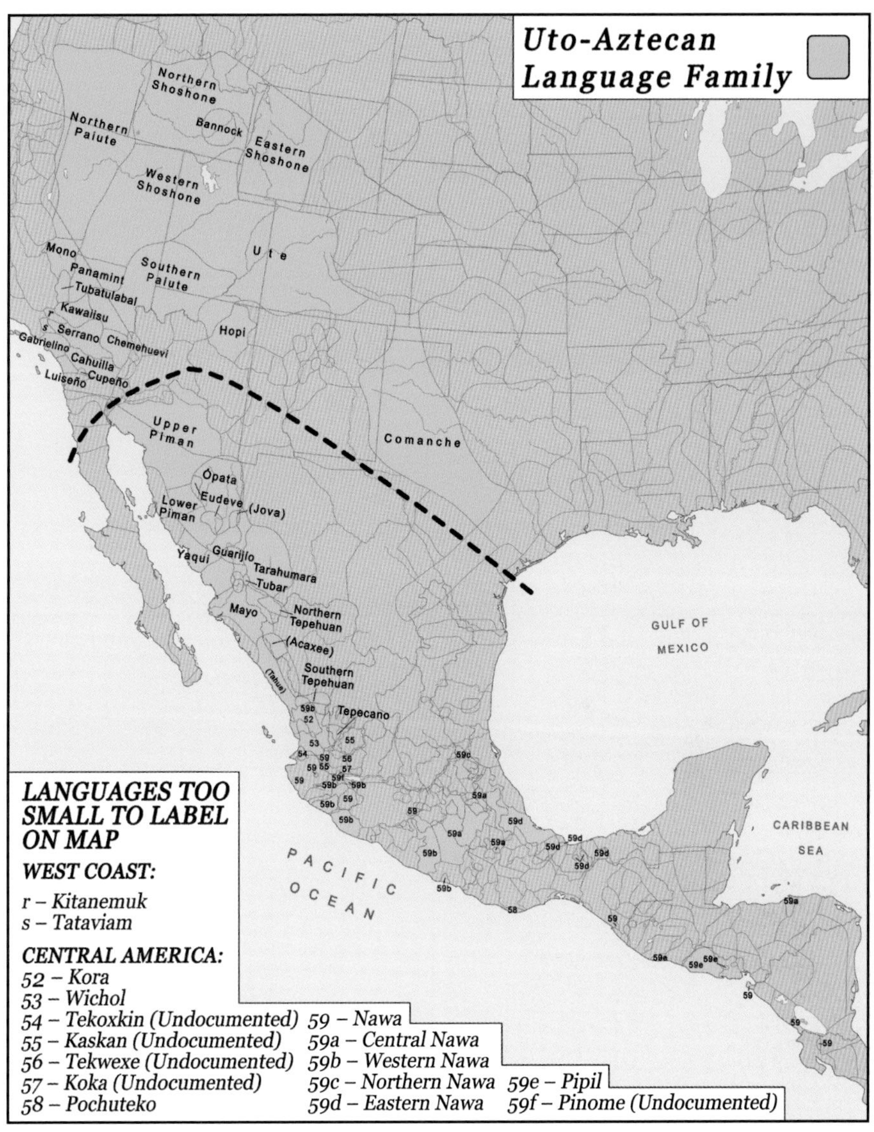


**Figure 89:** Contact-era geographic distribution of members of the Uto-Aztecan language family, including relatives in Central America. The dashed line represents a linguistic division between Northern and Southern Uto-Aztecan languages.

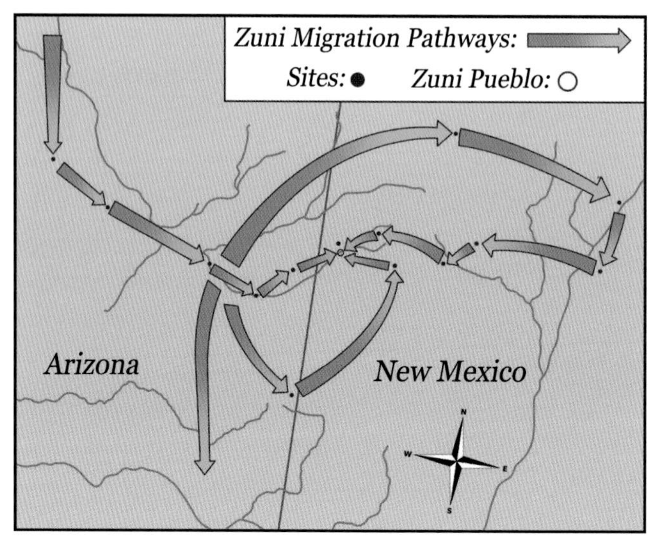


**Figure 90:** Map of ancestral Zuni migrations.

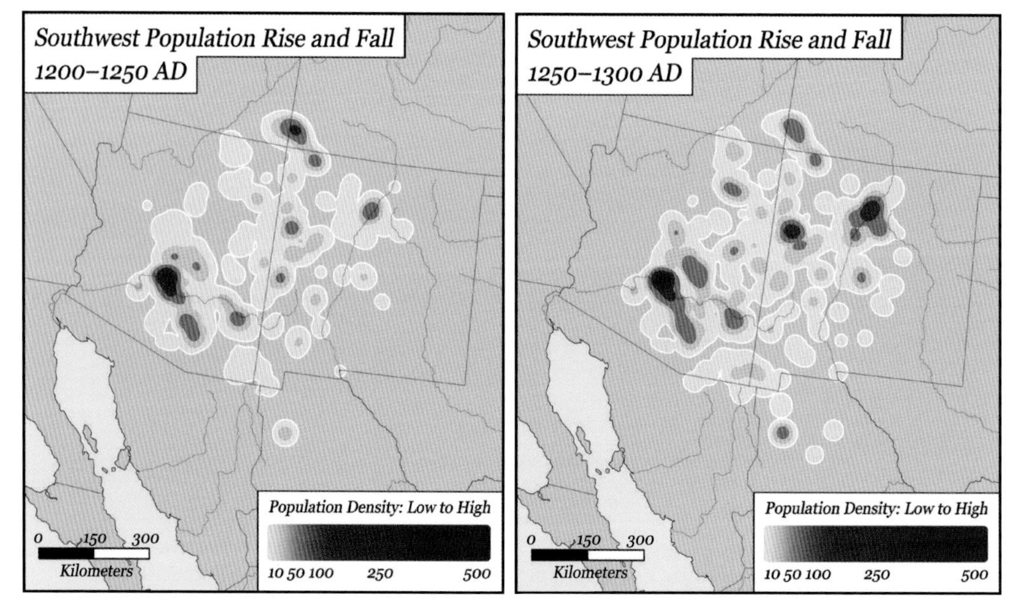


**Figure 91:** Map of changes in population size in the southwest in the final centuries before Contact. See additional population maps on the following page.

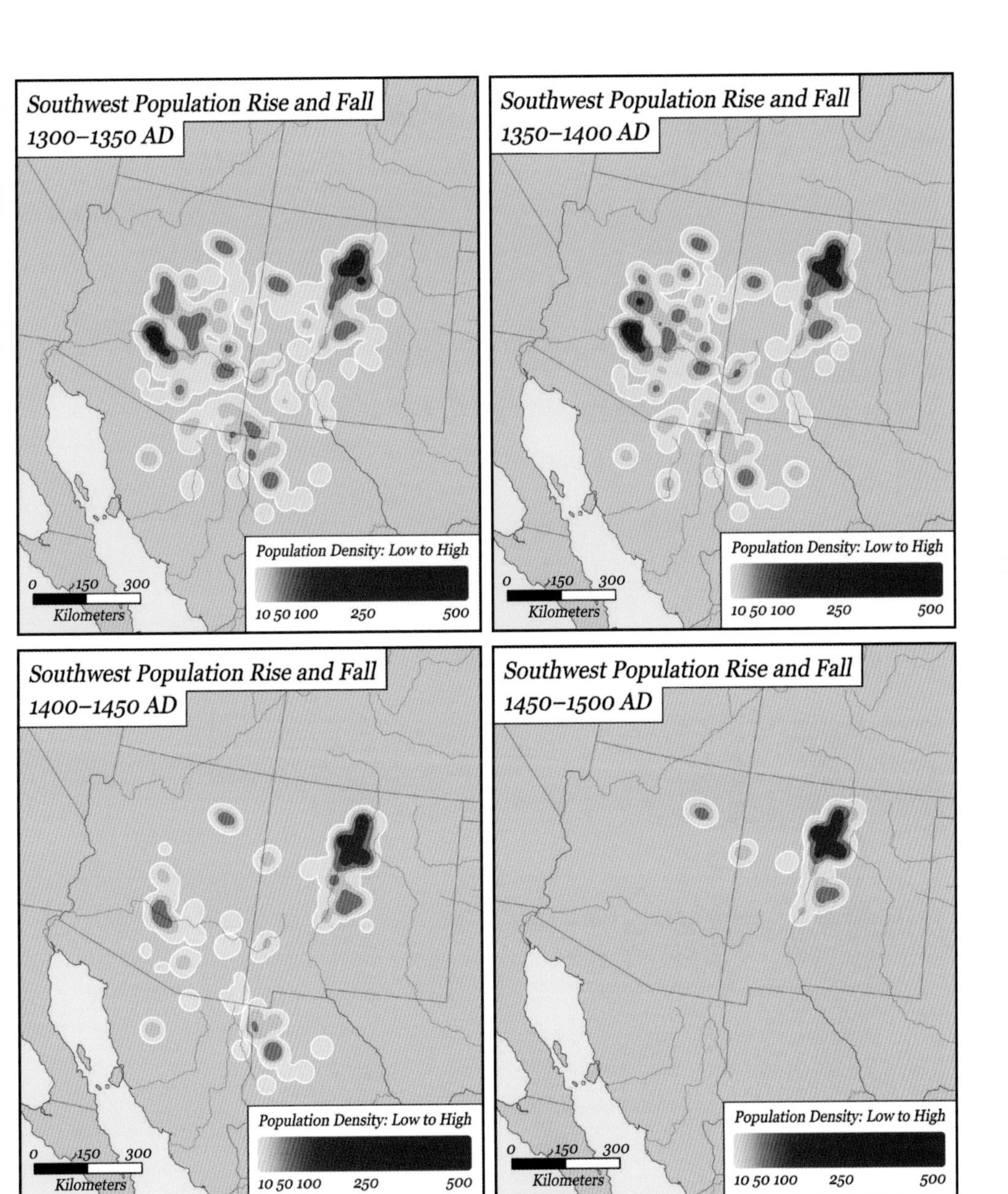

Southwest Population Rise and Fall
1300–1350 AD
Population Density: Low to High
10 50 100
250
500
0
150
300
Kilometers
Southwest Population Rise and Fall
1350–1400 AD
Population Density: Low to High
10 50 100
250
500
0
150
300
Kilometers
Southwest Population Rise and Fall
1400–1450 AD
Population Density: Low to High
10 50 100
250
500
0
150
300
Kilometers
Southwest Population Rise and Fall
1450–1500 AD
Population Density: Low to High
10 50 100
250
500
0
150
300
Kilometers

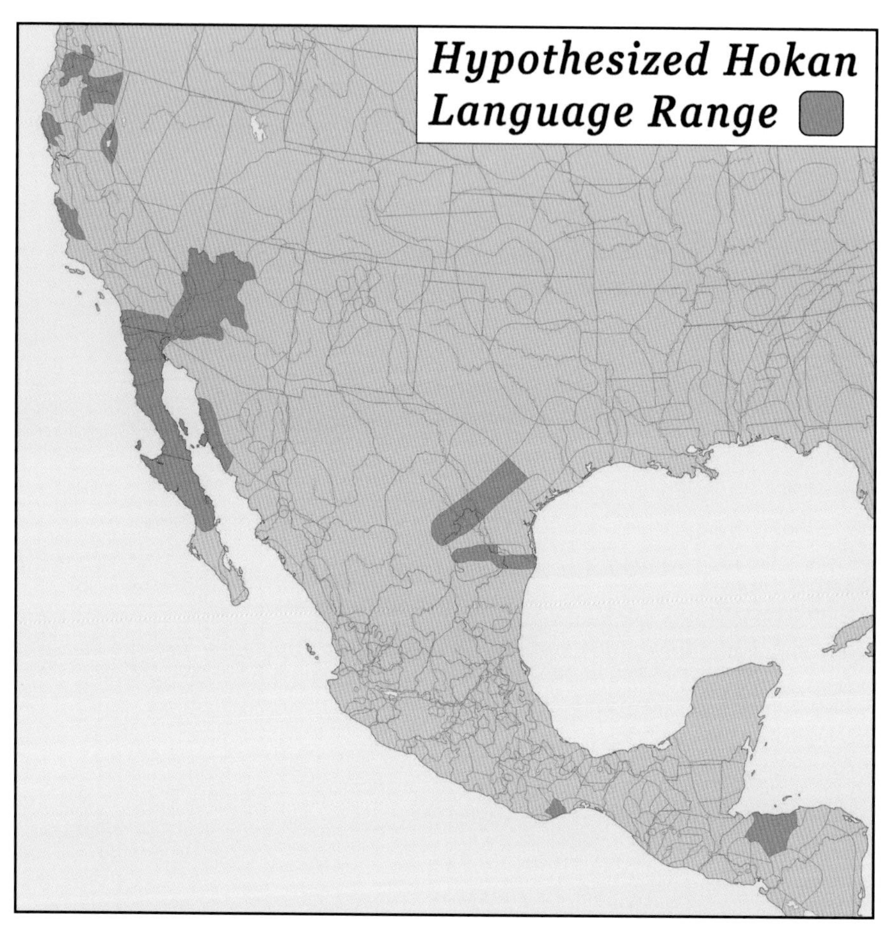


**Figure 92:** Map of the geographic extent of members of the hypothesized Hokan language family. Labels for individual language groups omitted. Compare to **online map** for labels.

**Figure 93:** Spruce Tree house at Mesa Verde National Monument in southwestern Colorado.

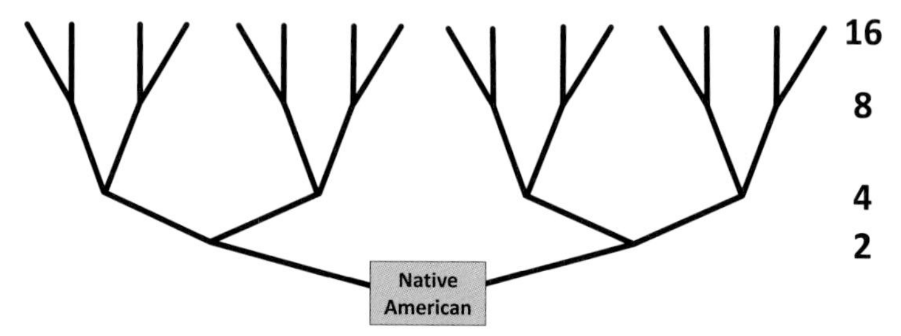


**Figure 94:** Example of how the number of ancestral branches doubles each generation as you go back in time.

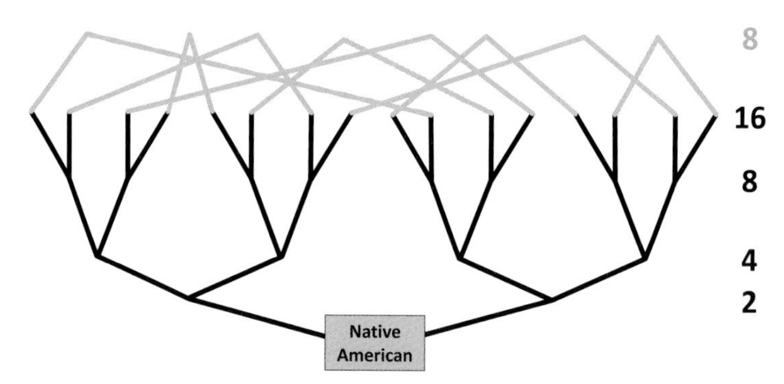


**Figure 95:** To reduce the number of ancestral branches to a reasonable amount, ancestral branches must eventually be connected to one another.

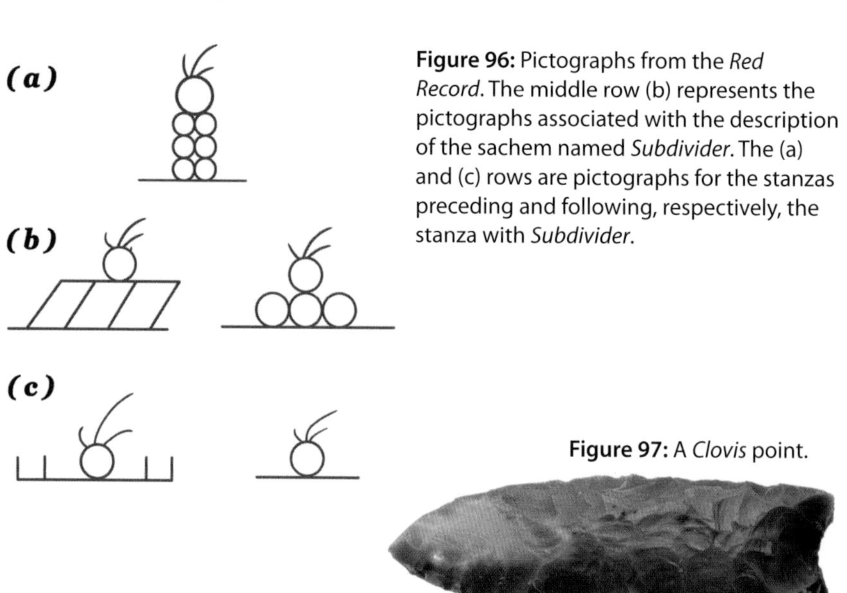


**Figure 96:** Pictographs from the *Red Record*. The middle row (b) represents the pictographs associated with the description of the sachem named *Subdivider*. The (a) and (c) rows are pictographs for the stanzas preceding and following, respectively, the stanza with *Subdivider*.

**Figure 97:** A *Clovis* point.

> originated from Mo-ning-wuna-kaun-ing (La Pointe), and the phrase is often used in their speeches to the whites that "Mo-ning-wuna-kaun-ing" is the spot on which the Ojibway tribe first grew, and like a tree it has spread its branches in every direction, in the bands that now occupy the vast extent of the Ojibway earth; and also that "it is the root from which all the far scattered villages of the tribe have sprung"...
>
> It is comparatively but a few generations back, that this tribe have been known by their present distinctive name of Ojibway. It is certainly not more than three centuries [i.e., late A.D. 1500s[13]], and in all probability much less. It is only within this term of time, that they have been disconnected as a distinct or separate tribe from the Ottaways ["Ottawa"] and Potta-wat-um-ies ["Potawatomi"]...
>
> The final separation of these three tribes took place at the Straits of Michilimacinac from natural causes.[14]

According to Warren, the timing of the Ojibwa migrations from the Atlantic

> in all likelihood, occupied nearly two centuries prior to their final occupation of the shores of Lake Superior.[15]

The latter event, again according to Warren and his Ojibwa contacts, could be assigned a rough calendar date:

> It is now three hundred and sixty years since the Ojibways first collected in one grand central town on the Island of La Pointe.[16]

Warren died at age 28 in A.D. 1853, but his *History of the Ojibway Nation* wasn't published until A.D. 1885. This would put the date of the arrival at La Pointe roughly in the early A.D. 1500s,[17] and the departure from the Atlantic around the early A.D. 1300s.[18] The details of

---

13 Book was published in 1885. A.D. 1885 - 300 = A.D. 1585.

14 Warren 1885, p. 80-81.

15 Warren 1885, p. 91.

16 Warren 1885, p. 90.

17 Date range of A.D. 1493 to A.D. 1525.

18 Date range of A.D. 1293 to A.D. 1325.

Warren's calculations show that these numbers were rough estimates, not precise calendar years.[19] Therefore, with a little adjustment, we can easily harmonize Warren's timeline with the *Red Record*'s implied date of origin for the Ojibwa, namely, the late A.D. 1300s (see above).

Why did the Ojibwa leave the Atlantic? Did someone else send them west? The answer is ambiguous:

> The original cause of their [i.e., the Ojibwa's] emigration from the shores of the Atlantic westward to the area of Lake Superior, is buried in uncertainty. If pressed or driven back by more powerful tribes, which is a most probable conjecture, they are not willing to acknowledge it.[20]

If I were in their shoes, I don't know that I would acknowledge it either. However, the archaeology of the Arctic puts a new angle on this question. In **chapter 5**, we speculated on who the Algic first encountered, not on the Atlantic, but on the Pacific—specifically up in Alaska. An Alaskan archaeological culture, the *Thule*, began migrating eastward out of Alaska around the same time that the Algic arrived. The Thule eventually reached what are now the Canadian provinces of Newfoundland and Labrador, which sits just to the north and east of where the St. Lawrence empties into the Atlantic (**Figure 66**). Here, in the east, the Thule seem to have replaced the prior culture, the *Dorset*.[21] I wonder if this sequence of events—and possible conflict—provoked the Ojibwa to turn around and head toward the Great Lakes.

Once on their way, it's hard not to wonder if the Ojibwa battled additional enemies. If we visualize their migrations on a map (**Figure 67**), the ancestral Ojibwa migrated through land that, at Contact, was Iroquois country (**Figure 67**).

What about pre-Contact? Among the northern Iroquoians (i.e., Iroquoians who are not Cherokees, see **Figure 67**), "'Iroquoian' cultural traits were…fully developed" at "the turn of the 14th century"[22]—or right around the time that the Ojibwa left the Atlantic. This date makes the northern Iroquois candidate foes for the Ojibwa.

19 Warren 1885, p. 90-91.
20 Warren 1885, p. 82.
21 Mason and Friesen 2017, esp. p. 144.
22 Birch 2015, p. 271.

Were they? Maybe. From the Iroquoian side of the equation,

> While there is some evidence for violence in the 13th and 14th centuries...conflict increased dramatically throughout Iroquoia during the mid-15th through early 16th century.[23]

By Warren's math, the migration through the northern Iroquoian areas happened in the A.D. 1300s and 1400s, and maybe into the early A.D. 1500s. This overlaps in part with the Iroquoian period of "some violence" in the 14th century. It definitely overlaps with the period of heavy violence of the 15th and early 16th centuries. Perhaps the Ojibwa were the cause.

I don't know for certain if the Ojibwa ever fought with the Iroquois. But the main body of Algonquians almost certainly did.

By the time the Algonquians crossed the Ohio and arrived on the eastern seaboard, the *Mahongwi*, which McCutchen translates as *Iroquois*, "were trembling."[24] The estimated date range for this event was mid to late A.D. 1400s. In other words, exactly when Iroquoian archaeological evidence reveals more conflict (see previous section).

Fights with the Iroquois—the *Mahongwi*—show up again near the end of the *Red Record* in the late A.D. 1500s.[25] This is about a century after the "Wolf People" separated from the remaining Algonquians. But the geography of the "Wolf People"—the Mohegans—as well as the remaining Delaware, at Contact made them neighbors to the Iroquois (**Figure 24**). And potential combatants.

Which means that, had battles with the Talamatan and/or Iroquois gone differently, or had events on the Atlantic played out another way for the Ojibwa, someone other than the Wampanoag would have feasted with the Pilgrims in the fall of A.D. 1621.

23 Birch 2015, p. 283.
24 McCutchen 1993, p. 128.
25 McCutchen 1993, p. 136.

# 9

# Escape
# A.D. 900s to Contact

I stood anxiously with a healthy distance between me and the fencing designed to prevent a fall of more than 33 stories down the light tan and black Arizona rocks of *Walnut Canyon* to the bottom 400 feet below. The fencing blocked only part of the canyon edge. The wind was relentless with gusts over 20 mph. Had my wide-brimmed hat not been tied under my chin, it would have quickly become part of the canyon ecosystem. I was afraid that, if I wasn't careful, I'd end up with a similar fate.

Prior to Walnut Canyon, the journeys of the Algic and my visits to the mounds of the Hopewell, Effigy, and Mississippians had taken me historically or physically through almost every part of what are now the U.S. and Canada—except for the southwest (**Figure 68**). *Surely the southwest had a story to tell.* So, on Saturday, June 1, 2024, I flew into Phoenix for a week's worth of visits to Native American sites in this region. On Monday, June 3, I was more than a two-hour drive north of Phoenix and less than 10 miles east of Flagstaff in Walnut Canyon National Monument.

Through this cavernous landscape (**Figure 69**), the blowing of the wind sounded less like angry howling and more like hollow whoosh-ing. I was alone. Though the early afternoon sun was bright with temperatures in the 80s, the atmosphere was almost haunting, which added to the trouble of steadying my nervous legs at this height, legs which were already unsteady from fateful events a few minutes earlier.

Prior to the trip, I knew from the academic literature that the Natchez empire had never reached into this part of the United States. No kingdom as large as the Mississippian empire brought all the southwest under its sway. Or was there more to the political history of the southwest? If the Natchez built the North American equivalent of the ancient Roman empire, did anyone out here construct an analogy to, say, the Holy Roman Empire of the Middle Ages? Walnut Canyon, among many other places, would cast these questions in a new light.

I also knew the era on which to focus. The southwest never saw a phenomenon like the Hopewell culture. Before the time of Christ, only mobile bands of hunters and gatherers roamed that area. Permanent villages didn't rise until the A.D. 500s to 700s,[1] and these were clusters of *pit houses*, the earliest type of home in this region, before the standard residential unit shifted to pueblo designs. The most dramatic episode in southwest history, and the main focus of my trip, was the time of the cliff dwellers.

Twenty minutes earlier, I had gained a poignant appreciation for their daily life-and-death decisions. I had attempted to descend the 273 steps down from the visitor center—perched comfortably at the end of the cliff about 500 yards distant from where I now stood—to the Island Trail below. I had gone down about 10 steps before I abruptly changed my mind. Too much blue sky rushing away behind the steps; not enough solid ground to catch any missteps. I had panicked at the vertigo and turned around.

I had seen that there was another trail—the Rim Trail. I had taken the latter to arrive at my current perch. But the Rim Trail had no cliff-side ruins. Boo.

From my position, I spotted some of the ancient brick and mortar homes nestled into the side of the mountain. The specific cliffs on which the Native Americans had chosen to erect their dwellings were found on a sort of peninsula, around which snaked an occasional stream at the base of the canyon. A spit of wall connected the peninsula back to

1 i.e., see Fagan 2019, p.212; see also Plog 2008, p. 56.

the main canyon wall, at the edge of which sat the visitor center. There was only one way to get to the ruins. Hence, the *Island* Trail.

Prior to the trip, I had dug into the histories of the indigenous peoples of the area. *Hopi, Zuni, Tewa, Tano, Tiwa, Acoma, Pima*, and other Pueblo peoples, together with the arriving Athabaskans—the Navajo and Apache—dwelt in the core of the Southwest at Contact (**Figure 70**). *Utes, Paiutes, Comanche*, and members of the *Cochimí-Yuman* language family, such as the *Havasupai, Hualapai, Yavapai*, and *Maricopa* lived along the periphery of the southwest (**Figure 70**).

Unfortunately, unlike the histories for the peoples elsewhere in the United States and Canada, the histories of the southwestern peoples were bound up in religious or seemingly symbolic language. I couldn't make heads or tails of them. I learned that the mainstream community had struggled with these same indigenous narratives for decades. Yes, there were exceptions to this rule. Yes, the Athabaskans had a history of migration into this region (see **chapter 5**). Yes, the Caddoans had come out of the southwest (see **chapter 7**). And, yes, by the end of this chapter, you'll see some of the tentative attempts to reconstruct the ambiguous southwestern migration narratives. But these exceptions aside, the history of the southwestern peoples remained a mystery. *There goes one of the richest sources of history for this region*, I thought—at least for the time being. To uncover the events in the southwest, I'd need to rely heavily on archaeology.

I didn't relish the thought. Archaeology, by nature, is ambiguous. Many competing narratives can explain a single observation. How to choose among them? Unlike the mainstream community, I had a rich context—the detailed history for other parts of North America, as per the preceding chapters—in hand. Maybe some parallels would spark an insight. Maybe a story would emerge. Maybe.

But I would have to visit the sites. Even the cliff dwellings.

I stood reconsidering my earlier decision, weighing the risks and rewards. I now saw that the steep slopes were not sheer drop-offs, but almost stair-step in appearance. Short sheer drop, then small landing, then another sheer drop, then another landing, etc. Perhaps the danger wasn't as great as I initially thought. Perhaps the terror would be less if I could just keep telling myself that my eyes were lying while descending 185 feet to the cliff dwellings. Perhaps.

◂◂ ▪ ▸▸

I hadn't expected a terrifying experience here. In retrospect, I suppose it's mind-numbingly obvious that Walnut *Canyon* would have breathtaking overlooks and trails. But I hadn't thought of the risks prior to arrival. I was more worried about the tour I had booked for Wednesday, June 5, in southwestern Colorado at *Mesa Verde National Park*, a well-known site for cliff dwellings.

One of the main tours there—of *Balcony House*—warned the ticket-buyer up front. "A tour of Balcony House is one of the most adventurous in the park and not for the faint-of-heart. You will scale the face of a cliff via several tall ladders, squeeze your way through a narrow tunnel on your hands and knees, and explore some of the same passages used over 800 years ago!"[2] Past visitors had uploaded videos of the entire tour to YouTube. They showed the final ascent out of the cliff dwelling—up a set of stairs carved into the cliff face. Stairs with just a netting or chain-link fencing strung beside you to keep you from falling to certain death on the rocks far below. *No thanks*, I thought.

Instead, I had booked a tour of *Cliff Palace*. The description didn't blare any headlines about fear of heights. The YouTube videos of tour-goers didn't seem as bad as Balcony House. But I was still worried.

I was not prepared to confront my fear of heights at Walnut Canyon.

I returned to the visitor center by the circuitous Rim Trail path that led past an ancient pit house and above-ground stone pueblo house. Maybe I was trying to buy myself more time to reconsider the perilous Island Trail. But visiting houses on the canyon rim was not the same as visiting cliff dwellings. There was no way around this fact.

So I re-entered the visitor center and asked the worker behind the counter about the Island Trail. She said she didn't get vertigo and wouldn't be much help. "Going down is optional, but returning is mandatory," she said with a smile, quoting the phrase from the Island Trail sign. I hesitated again.

---

2 https://www.recreation.gov/ticket/233362/ticket/500

And then I went. With fear and trepidation, I laser-focused my gaze on each step in front of me, gripping the railing or placing my hand on any available rocks to keep my balance. I kept reminding myself that the sudden lack of trees beyond the rail wasn't as bad as it looked. And I made it to the bottom, where the actual trail began in earnest.

I set out and soon discovered that the trail along the cliff dwellings had little to no guardrails. And it was one way. There was no going back.

On each trip I've taken to Native American ruins, I've diligently documented all that I saw with video and photos from my cell phone. I resumed the practice on the trail for a few cliff dwelling stops. Then the steep angle of the rapidly disappearing trees to my right became too much. *Forget the picture taking*, I thought. I just needed to get off the mountain.

In times past, I have been asked, "Do you fear falling? Or do you fear jumping?" *Yes*, I had answered. But now I had a third fear confronting me: The danger of fainting. Was I feeling light-headed? I made a conscious effort to breathe, regularly and deeply. If I didn't, who would catch me? Who would stop my unconscious body from tumbling from stair-step ledge to stair-step ledge until I reached the bottom? Again, I laser-focused my eyes on the path in front of me. My pace quickened.

I realized the path made a loop. But I didn't know how much farther I had to go. Eventually I rounded a bend. Maybe halfway home? No. I rounded another bend or two and recognized that I was on the side I had seen from the Rim Trail earlier. Now I was halfway. Maybe.

The trail kept going. Much as I wanted it to end, it wouldn't. It was hard to keep my eyes from sweeping to my right and the vast empty space beside me. And then, finally, I reached the staircase back up.

*Keep breathing*, I thought. But now it was harder. It wasn't just going up. I was climbing stairs at 6,700 feet above sea level. Would I faint for lack of oxygen? Was I feeling light-headed again? Did I need to pause and catch my breath? When would I reach the top?

I'm guessing I looked spooked by the time I arrived back at the visitor center. The same worker from earlier caught a glimpse of me as she was entering an office. "How was it?" "*It scared the mess out of me*!" I said. I guess to encourage me, she said, "You deserve a medal!" I wonder if my face was green.

◀◀ ▪ ▶▶

Who were the ancient inhabitants of Walnut Canyon hiding from? Who were they scared of? I didn't see any other explanation for their dramatic abodes than defense against enemies.

"There's no evidence of that," announced the park ranger when I recited these musings to her.

*What?* Apparently, the skeletons recovered from these sites showed no signs of conflict with invading forces. Still, I found her conclusion hard to fathom.

Earlier in the day, I had visited *Montezuma Castle National Monument* (**Figure 71**). Another cliffside dwelling, it sits one hour south of Walnut Canyon. Yet it's almost 4,000 feet lower in elevation[3] in the valley of the Verde River.

Unlike Walnut Canyon, the visitor no longer is allowed to walk astride the ruins. I'm not even sure how one would do so. The stone houses were huddled about 100 feet above the valley floor[4] up the red and white walls of a nearly sheer cliff face. It was awe-inspiring to see. And much less terrifying to experience.

But the implications seemed to be the same as at Walnut Canyon. Surely the builders were trying to escape someone or something. Why erect a structure accessible only via a series of ladders? Why risk the lives of your children day in and day out? As a parent, I can't imagine the thought of having to pick up my lifeless child's body from a fall from that height. I can't imagine putting them in that danger under any circumstance—except one: to escape a greater threat.

Before visiting Montezuma Castle, my first stop of the day had been at *Tuzigoot National Monument* (**Figure 72**), a 40-minute drive northwest of Montezuma Castle. No cliff dwellings there, but Tuzigoot did contain the ruins of 110 rooms, all built of stone and mortar. It was

3 https://en-us.topographic-map.com/map-8f3/Arizona/?center=34.61159%2C-111.83096&zoom=15

4 As per the park brochure.

constructed in stages, from A.D. 1050 to 1380.[5] One room, which I was allowed to enter, still had the roof and original wooden beams visible in the ceiling.

Even without cliff dwellings, the defensive positioning of Tuzigoot was hard to miss. The area surrounding the pueblos was shaped almost like a bowl. Mesas or mountains surrounded the valley on three sides. The pueblo ruins sat off-center, but on a hill that rose 120 feet above the valley floor.[6] The closest approach was a mesa about 0.4 miles away which guards the view to places beyond. In other directions from where I was standing on the summit, I could see for 4-5 miles. From this vantage point, the approach of any invader would have been obvious.

Attacking Tuzigoot would have been challenging. Sure, you couldn't have defended Tuzigoot the way you could have defended Walnut Canyon—by shoving attackers off the cliff—yet once an enemy decided to advance on Tuzigoot, even from a mesa top, they would still have had to drop into the valley before ascending and fighting uphill.

Once I learned the dates for the two earlier sites, the intrigue surrounding all three locations grew. Montezuma Castle was inhabited from the A.D. 1130s to the 1400s.[7] Native Americans in Walnut Canyon dwelt there from A.D. 1100 to 1250.[8] Interesting. All three sites were contemporaries in time. And all three were seemingly concerned with defensive positioning.

My memories from the experience from the day prior (Sunday, June 2) added to the emerging pattern. In the morning, I had driven up from Flagstaff to Tuba City to speak at a Navajo church. On the way back in the afternoon, I had visited *Wupatki National Monument* (**Figure 73**), about a one-hour drive[9] northeast of Flagstaff. Unlike Walnut Canyon, Wupatki sat in the 4,700-foot to 5,400-foot elevation range, depending on where you were in the park.[10] Turning into the park from highway 89, the entrance to Wupatki sits in the middle of

5 As per the signage at the site.
6 As per the park brochure.
7 As per the signage at the site.
8 As per the signage at the site.
9 To the Wupatki visitor center.
10 https://en-us.topographic-map.com/map-8f3/Arizona/?center=35.52239%2C-111.36982&zoom=16
https://en-us.topographic-map.com/map-8f3/Arizona/?center=35.56842%2C-111.47337&zoom=16

wide-open grasslands. By the time I dropped into the area with the visitor center and main pueblo, the landscape was dry and full of scrub, red earth, and red stone.

But the views were stunning. From the visitor center, which sits just above the main 100-room red brick pueblo, I could see 60 miles east to the Hopi Buttes. The main vista opened slightly northeast to the multi-color walls of an escarpment on the Navajo reservation—20 miles away. The original builders of the pueblo positioned themselves by a spring, in a slight depression. But I spotted another pueblo, which was not open to the public, on a nearby mesa top with a line of site to the main pueblo.

Almost three miles due east of *Wupatki* pueblo sat *Wukoki* pueblo (**Figure 74**). The Native Americans built it on top of a boulder. A set of modern steps took me up into what felt like a lookout tower. The top of the red boulder was broad enough that the entire structure felt like it could have functioned as a combined multi-room residential unit and advanced warning fortification. In terms of size, it was several times smaller than the main Wupatki pueblo. Line-of-site connected me back to the main visitor center. It was hard to conclude that Wukoki was anything other than a defensive outpost for Wupatki.

Driving back northwest toward the Wupatki entrance, I stopped by another site called the *Citadel* (**Figure 75**). The name fit perfectly. It was built on an isolated hilltop in the midst of the grassy landscape.

On the hilltop, the crumbling red and black brick remains of a multi-room structure were visible. Unlike the main Wupatki and Wukoki pueblos, no high-walled structures endured at the Citadel. At the Wupatki and Wukoki ruins, I could walk into rooms where the walls still rose high above me and, for a few minutes, would take me back in time to taste life in northern Native Arizona.

One mile due north across the grasslands sat the walled remains of small pueblos in Box Canyon. Many of the brick walls and rooms of this cluster of homes had resisted the ravages of time. I could walk through a few. Though the canyon wasn't more than one to three stories high—not high enough to give me vertigo—the ancient builders still took pains to position their homes exactly at the cliff edge (**Figure 76**). Pushing an enemy over the side wouldn't have been fatal. But it was obvious that the options for attack were immediately limited by the residential positioning. Line of site connected the homes to one another and, for at least one or two of the pueblos, to the Citadel. Again, it was hard to imagine the layout to be anything other than preoccupied with advanced warning and defense against enemies.

Wupatki was inhabited beginning in the A.D. 1100s—again, contemporary with the other defensive sites that I had witnessed.[11]

Surely *defense* explained the perilous nature of so many of these locations.

On Wednesday, June 5, two days after my scare at Walnut Canyon, my experiences gelled my developing conclusions more firmly in my mind. Yes, it was while exploring Cliff Palace and other ruins at Mesa Verde National Park. No, it wasn't because I fell off the edge or had a nervous breakdown. On the tour of Cliff Palace, I found my fears of vertigo happily unfulfilled. Sure, the terror of staring down a steep drop kept me from lingering long at any of the overlooks, but the stairs down to the site, as well as the gentler slope that fell away from the 150-room compound, didn't have nearly the same effect on me as the trail outside of Flagstaff.

The drive up to Mesa Verde[12] was another story. The site rises almost like a castle or tower from the surrounding valley floor. Some of the highest points reach 8,000 feet above sea level. Low points rise to only 5,000 feet. The winding two-lane road zig-zags the visitor back and forth along the edge, with guardrails in the most dangerous positions. And none in many other positions, positions that produce sweaty palms and tingling legs. I was happy when I finally reached the main parking lot on Chapin Mesa, where the Archaeological Museum sits. Happy, that is, until I realized I'd have to repeat the drive at the end of the day to get back down to the valley floor.

Cliff Palace and Balcony House are just two of the many sites to see in the Mesa Verde park. After my tour, I re-drove the loop roads and stopped at each overlook and point of interest. It was then that

11 As per the booklet and NPS website.

12 Technically, Mesa Verde is a *cuesta*, because it "is inclined slightly to the south at about a seven-degree angle" (https://www.nps.gov/meve/learn/nature/index.htm).

the scale of the cliff dwellings became more apparent. As spectacular as Cliff Palace can be (**Figure 77**), I discovered that it is but one of a whole series of dwellings that line the cliff faces to its north and south.

Back on the Cliff Palace tour, the park ranger was quite happy to entertain *defense* as an explanation for these perilous dwellings. *Good*, I thought.

By the end of my trip, the evidence had (to me) become relentless. For the most extreme cases—Walnut Canyon, Mesa Verde—I noticed numerous positions in the surrounding area where the Native Americans could have built their houses but didn't. Instead, they chose the most dangerous positions for their residences. Even at Wupatki, it was clear that the builders chose the exact edge of Box Canyon for their home, not 20 feet away in a less treacherous location. At Walnut Canyon, the big aha moment came as I was driving away on I-40 east toward Gallup, New Mexico. Just outside of Flagstaff, I crossed a bridge with a sign announcing *Walnut Canyon*. Yet the drop to my right was maybe 20-30 feet. Hardly a *canyon*. Why didn't the builders erect their homes here, in a nice safe location without the daily danger of falling to your death? *Because there was a greater danger of attack from enemies*, I thought. It didn't seem like there was a better explanation.

I also realized I had witnessed a general pattern. As the crow flies, the two most-distant sites I had visited, Tuzigoot and Cliff Palace, were 250 miles apart. Thousands of feet of elevation separated them from one another, as well as from the Wupatki, Montezuma Castle, and Walnut Canyon locations. These sites also spanned a variety of ecosystems. Yet, despite these diverse settings, all were preoccupied with one goal: defense against enemies.

All were also contemporaries in time. Occupation of the cliffside dwellings at Mesa Verde began in the late A.D. 1100s.[13] These dates were in line with the ones for Walnut Canyon, Montezuma Castle, Tuzigoot, and Wupatki (see above).

Common purpose, common time period. I wondered if they all feared a common foe.

If only I could find out who.

13 As per the booklet.

◂◂ ▪ ▸▸

The day after Mesa Verde, I found myself in northwestern New Mexico, an hour-and-a-half drive from any semblance of civilization. The closest hotel to *Chaco Canyon National Historic Park* that I could find was in Farmington. I had chosen Farmington after calling the park the week prior regarding the state of the roads, as per the recommendation on the website. It's possible to stay south of the park in Gallup, but the staff member on the phone urged against it. He advised the approach from the north as the southern approach was extremely rough and, at times, impassable. Even then, the website announced that 13 miles of the northern road into Chaco were unpaved.

I discovered exactly what they meant. A good chunk of the "unpaved" section was gravel. Then it turned to packed dirt. I navigated slowly as I bumped along in a low clearance rental car. *Do they want people to visit?* I thought. If this was the state of the preferred road, I couldn't imagine how bad the south approach was.

At one point, I was struck by the utter remoteness of the location. Part of the drive winds past (through?) Native reservation land, which tends to be rural anyway. But still, even though I had the windows up and the A/C running, it immediately imbued me with a hushed sense, an atmosphere of isolation and solitude.

*Why build here?* I wondered. And I kept pondering throughout the day.

The sun was hot. At one point, my car temperature showed triple digits. *Chaco Wash* winds through the valley between the mesa tops. Along with the runoffs from the mesas, the wash supplied water to the ancient residents, but the site as a whole looked parched. The cliff walls were tan; the soil was dusty to sandy. The plant life was, at best, scrub. No cacti, but a swimming pool seemed like a minimum prerequisite for anyone wanting to call this place home.

On the drive in, I saw what amounted to mostly flat terrain, with some rolling hills. The road gently dropped me into the canyon, though once I was there, the location gave off hints of the mesas from Mesa Verde. The descent from mesa top to valley was sheer cliff face, and the height was 350 feet or less. But the canyons themselves were fairly flat and very wide (**Figure 78**). Plenty of room for farming. Kind of cozy, aside from the semi-desert atmosphere. It felt less haunting than Walnut Canyon and definitely not as dangerous as Mesa Verde. Maybe because I was on the ground.

Then again, I briefly flirted with the idea of taking one of the many hiking trails, this one from the Canyon floor up to *Pueblo Alto*, one of the mesa top sites. No guardrails exist at the top. But the rough uncut stairstep approach up a crack in one of the rocks felt safe enough. I almost took it. But it was 2:45 p.m., and I was alone with little to no cell phone service. I even had to sign the book at the trailhead, the ranger had told me back in the visitor center, so that when they made the rounds at 9:00 p.m., they would know if someone had gotten lost or stuck in the wilderness. *Great*, I thought, *if a rattlesnake bites me, no one will find me for hours. Better wait until I have a hiking partner.*

Despite the heat and barrenness, the majestic setting for the canyon emerged as the day went on. Chacoans were known for their *Great Houses*—planned, multi-story compounds often with splendid masonry work. In other words, Chacoans built houses that suggested organized labor. Organized labor suggested hierarchy.

Chaco contained several Great Houses that were spread out along the canyon floor and up on the mesas. From *Wijiji* in the southeastern end to *Peñasco Blanco* in the northwestern tip, the distance spans seven miles. A loop road took me around to a portion of the Great Houses. Hiking trails would have taken me to the rest.

Thankfully, the most famous of the Great Houses, *Pueblo Bonito* (**Figure 79**), was on the main driving loop. The booklet for the self-guided tour took me through the ruins and even some of the rooms with still-standing multi-story walls and original wooden beams. One estimate put the population size for this 600-room complex around 1,000.[14] The exact population number, though, is debated. Some estimates go as high as 2,000; others as low as 100.[15]

This narrowly focused issue of population count is bound up in the larger questions surrounding Pueblo Bonito: What was the function of this magnificent site? Was it simply a large residential compound? Was

14 Fagan 2019, p. 225.

15 Fagan 2019, p. 225; Plog, S., Heitman, C.C., and Watson, A.S., "Key Dimensions of the Cultural Trajectories of Chaco Canyon," p. 289, in: Mills and Fowles 2017.

it purely religious, a pilgrimage destination used infrequently during the year? If so, I suppose this might explain its remoteness. Was it a royal residence, home to only a select few? The latter possibility grew on me as I walked and re-walked through the ruins.

At one point, I stood outside along the long flat wall of the D-shaped structure and re-entered one of the few entrances. In its day, Pueblo Bonito was restricted access; I wanted to relive what it may have been like for someone on the outside to be summoned in. As soon as I stepped into the main plaza, I noticed how the rounded shape created an almost amphitheater-like sensation. Before me about 80 yards distant, the still-standing back walls of multiple stories rose above my head. It must have been awe-inspiring when Pueblo Bonito was still inhabited.

But was it a royal seat? At least one of the burials in Pueblo Bonito suggested high status. I had little trouble imagining Pueblo Bonito as the seat of power, and the other Great Houses of Chaco Canyon as residences for subservient governors and other lesser officials.

Later in the loop drive, one of the stops took me on a brief trail around a number of small pueblo sites—only 10 to 20 rooms in size. In other words, *not* Great Houses. Interesting. A whole hierarchy of residences existed in a canyon that, at its height, may have been home for an estimated 5,000 people.[16] The *royal residence* hypothesis for the Great Houses seemed to fit.

Just 350 yards to the northwest of Pueblo Bonito, I visited *Pueblo del Arroyo*. Looking southwest, I was stunned by the natural setting. I saw a break in the mesas connected by flat land, the *South Gap*. My astonishment wasn't because of how the Gap revealed the function (or lack thereof) for Chaco Canyon. To be sure, the South Gap exposed the poorly defensible nature of the Great House locations. Rather, my wonder stemmed from the beauty of the view. Some kings roll out a red carpet to heighten the drama of anyone who wishes to approach their throne. The South Gap was around 400 yards wide with nothing but flat land stretching beyond as far as I could see. It was guarded by the mesa pillars. In other words, it was a runway of unequalled proportions.

16 Fagan 2019, p. 225-226.

Was Chaco a destination? The builders seemed to try to make it so. One of the most arresting, but only occasionally visible, aspects of Chaco culture was their road system. I got a glimpse on the loop drive back to the visitor center. A pull off let me view from a distance a staircase carved into the cliff face. When the Chacoans wanted to go somewhere, they paid little attention to natural obstacles in their path. But it wasn't just the audacity of the travel network that suggested importance for Chaco. It was also the sheer distance (**Figure 80**):

> Perhaps as many as 400 miles (650 km) of unpaved ancient trackways link Chaco to an intricate web with over thirty outlying settlements. The "roads" are up to 40 ft (12 m) wide and were cut a few inches into the soil, or marked by low banks or stone walls. Sometimes the road markers simply cleared the vegetation and loose soil or stones from the pathway, lining some segments with boulders. The roads run straight for long distances, in one instance as far as 60 miles (95 km).[17]

If nothing else, this brings to mind the fabled road system of the Hopewell of the East. Yet Chaco didn't seem to be just one of many links in this network. The roads suggested Chaco as the center, the bullseye (**Figure 80**), the destination of all other roads.

And it was indeed a network. The Great House template was replicated near and far. I had seen this firsthand two days prior when I visited *Aztec Ruins* about 50 miles as the crow flies to the north of Chaco. The same multi-story structure (**Figure 81**) and general site plan, along with careful masonry work (**Figure 82**), showed up here as well. At 400 rooms, Aztec Ruins was smaller than Pueblo Bonito. But I nevertheless found it impressive. In addition, while I wasn't allowed to walk up to it, the self-guided tour booklet mentioned that one of the Chacoan roads passed right through the site.

Even farther north at Mesa Verde, I saw the *Sun Temple* and its Chacoan design. Almost 40 miles northwest of there, in the *Canyon of the Ancients*, I witnessed a similar phenomenon at *Lowry Pueblo*. If we include the most distant sites with Chaco links, the Chacoan network reached 150 miles distant from the center (**Figure 83**).

---

17 Fagan 2019, p. 229.

If you're beginning to wonder if this sounds like an empire, that makes two of us. If you're also beginning to wonder if this specifically paralleled the Natchez empire from previous chapters, welcome to the club! In fact, building at Pueblo Bonito started in the A.D. 800s—close to the time that the Natchez commenced at Cahokia. Both groups saw their cultures replicated and multiplied across a wide stretch of land. For the Cahokians, rather than spread radially from a bullseye center, they spread eastward (recall **Figure 52**). From Cahokia to the South Carolina coast, it's about 700 miles. The widest diameter for the Chacoan network was smaller (**Figure 83**)—about 300 miles (480-kilometer diameter, 240-kilometer radius)—but still significant. Intriguing similarities.

One of the biggest questions about Chaco relates to its origins. The Natchez claimed origins in Mexico. Their flat-topped mounds emphasized the link. Chaco had no such mounds. But Chaco's trade network was replete with Mesoamerican items—chocolate, scarlet macaws, copper bells, seashells.[18]

Given contemporary events in Mesoamerica, it wouldn't surprise me in the least if Chaco was built by Mesoamerican immigrants. Prior to the A.D. 900s, Central America was in tumult. In the lowlands of Guatemala, the Classic Maya were just realizing their catastrophic end (see **chapter 3**). Mexican Mesoamerica also underwent a dramatic shift. In Central Mexico:

> Many of the cities that had forged a post-Teotihuacan order went into their own decline and eclipse. Much of the West Mexican area…had collapsed by this date. … As at the fall of Teotihuacan, there must have been much movement of peoples, shifting alliances, and a certain amount of social chaos.[19]

Out of this chaos arose the Natchez. Their ancestors migrated northeast out of Mexico. It doesn't take much of a mental leap to wonder if another Mesoamerican people went northwest to Chaco.

So far so good. Maybe Chaco was the empire from which so many surrounding Southwestern peoples were hiding.

18 As per signage at the visitor center.
19 Coe and Koontz 2013, p. 156.

Except that the Chaco Canyon civilization fell in the early A.D. 1100s, right as the defensive structures and cliff dwellings were gaining steam. The fall was sudden. Research has shown that expansions had been planned for Pueblo Bonito, yet, mysteriously, were never completed.[20]

Whoever the cliff dwellers were hiding from, it wasn't the Chacoans.

◂◂ ▪ ▸▸

So far in this chapter, I've told where the refugees fled *to*. I haven't told you where they fled *from*. Answers to the latter suggest geographic clues to the source of the enemy that the cliff-dwellers faced.

Let's line up and map out the multiple events that coincided with the adoption of defensive locations. One of the first events was not defensive construction, but a fall due, perhaps, to the lack thereof—the collapse of Chaco Canyon in the early A.D. 1100s. Directionally, Chaco was to the south of Mesa Verde.

Another region of the southwest, one which I haven't mentioned yet, is the *Mogollon* cultural area (**Figure 84**). It's to the south of Chaco Canyon. The *Mimbres Classic period* terminated at the same time as Chaco. The end was "marked by the depopulation of most settlements and considerable decline in population in the region due to emigration."[21]

Around the same time that Chaco fell, a region to the northwest of Chaco lost its population. Quantitative displays of the population rises and falls make this clear. Plots for Chaco and *Black Mesa* are strongly correlated (**Figure 85**).

As these three locations were losing their populations, Mesa Verde was gaining residents. Mesa Verde would eventually house some of the most defensively minded structures in the southwest. Geographically, Mesa Verde was to the north of Chaco and of the Mogollon area and to the northeast of Black Mesa (**Figure 86**). It's almost as if the refugees from Chaco, the Mogollon region, and Black Mesa fled to Mesa Verde.

---

20 As per signage in the visitor center.

21 Nelson, M.C. and Gilman, P.A, "Mimbres Archaeology," p. 270, in: Mills and Fowles 2017.

A movement from south/southwest to north/northeast suggests an enemy in the south/southwest.

Incidentally, the south/southwest is also the direction in which a resident of Mesa Verde would have to travel to get to Walnut Canyon, Tuzigoot, Montezuma Castle, and Wupatki. I'm not suggesting that the latter were the aggressors. Rather, that all of these defensively positioned sites formed a line that pointed from a south/southwest direction to a north/northeast location. In other words, they tell a consistent story.

In between these sites sit the Hopi mesas (**Figure 86**). During this period, the mesas saw significant population growth.[22] Yet the pueblo locations were defensively positioned,[23] and Hopi migration histories describe the conglomeration of peoples from very diverse locations in the southwest (**Figure 87**).[24] This suggests that the sites functioned similar to Mesa Verde—huddles for refugees seeking to escape some threat.

By process of elimination, only one corner of the southwest remains as a candidate for the menace: the southwestern-most corner (**Figure 86**).

◂◂ ▪ ▸▸

When I flew into Phoenix on June 1, I was in the candidate southern/southwestern region. My first two stops were *Mesa Grande* and *Pueblo Grande/S'edav Va'aki*. Both were sites of cultures that built platform mounds. At both sites construction on the large platform mounds began in the A.D. 1100s.[25]

When I visited that morning, Mesa Grande was closed. I guess when it gets too hot in Phoenix, outdoor activities stop. Despite being restricted to viewing from outside the fence, I still got hints of the sensations I experienced in the east. Namely, the moundbuilders

22 Bernardini, W., Peeples, M.A., Kuwanwisiwma, L., and Schachner, G., "Reconstructing Population in the Context of Migration," p. 141, in: Bernardini et al. 2023.

23 Bernardini, W., Lomayestewa, L.W., and Schachner, G., "*Wukokiikiqö:* Large Villages," p. 205-207, 217-223, in: Bernardini et al. 2023.

24 Hopkins, M.P., Kuwanwisiwma, L., Koyiyumptewa, S.B., and Bernardini, W., "Hopi Perspectives on History," p. 15-19, in: Bernardini et al. 2023.

25 For Mesa Grande, as per https://www.arizonamuseumofnaturalhistory.org/plan-a-visit/mesa-grande.
For S'edav Va'aki, as per signage at the site.

were *up there* and I was *down here*. The same power flex as with the Mississippians. At S'edav Va'aki, not only was the park and museum open, but I also got to walk around and on top of the mound. Again, same sensation.

Neither of the mounds were very tall. The base was about one story; another story was built on top of it. Yet the feeling of hierarchy remained.

Both Mesa Grande and S'edav Va'aki belonged to the larger cultural classification of *Hohokam*. The Hohokam culture area included south-central Arizona, down to the modern Mexico border (**Figure 84**). Or, to put it bluntly, the south/southwesterly direction from all the locations of all the defensive actions in the north.

Where did the Hohokam come from? From even farther south in … Mexico? Mainstream science tends to say *no*.

More accurately, mainstream science has answered both in the negative and the affirmative, depending on which period of literature you consult. Currently in vogue is the position that Hohokam rose at the hands of indigenous southwesterners, not Mesoamerican migrants.

We've encountered this type of mainstream science posture before. Recall that, in the east, mainstream views do *not* attribute Mississippian culture to Mesoamerican immigrants. Yet we find archaeological hints of the link—platform mounds, a diet involving significant amounts of maize, a hierarchical society. More importantly, we have the Natchez history explicitly describing an origin in Mexico.

Southwest of the southwest, we don't have anything like the Natchez history. But we have even more archaeological evidence for Mesoamerican links—and for a longer period of time. Maize—the classic Mesoamerican crop—appeared in the southwest in the early B.C. era. Later in pre-Columbian history, but before the period of platform mounds, the Hohokam culture was characterized by ballcourts. Ballcourts were the setting for a strongly Mesoamerican game that never showed up in the North American east. And then the ballcourt era was followed by the platform mound-building era.

In short, in the east, we have a major group of people (Natchez/Muskogean) with explicit origins in Mexico. Yet this region has less archaeological evidence for this fact than the southwest does. It would seem, then, that the Hohokam also came from Mexico.

In fact, the Hohokam may have arisen as a result of multiple migrations from Mexico. Surely a Mesoamerican group was responsible for the rise of the Hohokam ballcourt culture. When construction began on the platform mounds, the ballcourt system disappeared. This seems consistent with a new Mesoamerican group arriving and taking over.

Incidentally, Wupatki, the site north of Flagstaff with defensive positioning, has a ballcourt (**Figure 88**) but no platform mound. Given their typical association with Hohokam (**Figure 84**), I was surprised to find one that far north. I asked the park ranger, and she said it was the most northerly one. Makes me wonder if Wupatki represented a group of refugees from earlier Hohokam populations—populations who fled the Phoenix area once the platform mound-building Hohokam arrived.

Together, these observations point the finger towards the platform mound-building Hohokam as the cause of so much defensive angst in the greater southwest.

◀◀ ▪ ▶▶

Who built the Hohokam platform mounds? Unfortunately, at Contact, Hohokam no longer existed. Shortly before European Contact, this whole region underwent a collapse.[26]

At Contact, the formerly Hohokam areas belonged to the Pima (compare **Figure 70** to **Figure 84**). Linguistically, the Piman languages are part of the Uto-Aztecan family (**Figure 89**), and specifically the southern Uto-Aztecan branch. The members of the latter group—the relatives of Piman—are Mexican and Mesoamerican.

At first pass, this might suggest that the Pima themselves built the Hohokam mounds. Perhaps the Pima were the enemies from which the northerners tried to escape. But Pima indigenous history does not recount the destruction of the Pima Hohokam society at the hands of outsiders. It describes how the Pima were the ones doing the destroying. Their history specifically mentions[27] the destruction of one

26 Plog 2008, p. 179.
27 Russell 1975, p. 227.

of the largest Hohokam sites, *Casa Grande*, which sits around 40 miles southeast of Phoenix.

Upon reflection, this makes sense. Consider the Natchez. They once ruled an empire from Cahokia. A couple of centuries later at Contact, the Natchez were living about 500 miles south of their former home (**Figure 50; see also online map**[28]). No surprise for the losers of a battle; they had to give up their former domain. I'd expect a similar phenomenon to play out in the Hohokam area. Since the Hohokam were defeated pre-Contact, I'd expect the Contact-era residents to be the victors, not the former rulers.

Naturally, this raises the question of who exactly these former rulers were.

Indigenous histories from the southwest don't (yet) positively identify the builders of the Hohokam platform mounds. But they do rule out some candidates. Together with archaeology, they take us closer to the identity of the biggest threat in the pre-Columbian southwest.

Let's start the process of elimination in the bullseye center of the southwest and then move outward. At Contact, the Zuni were near the bullseye (**Figure 70**). Mainstream science has reconstructed the history of Zuni migrations (**Figure 90**).[29] While their account links to Mexico, it suggests a movement *into* Mexico, not out of it. In addition, the primary direction of Zuni movement is from the northwest, not the southwest. The Zuni seem like poor candidates for the identity of the Hohokam.

At Contact, the Pueblo neighbors of the Zuni were members of two language families, the Keresan and the Tanoan branch of the Kiowa-Tanoan (**Figure 70**). Both were in the vicinity of the Rio Grande. Both Tanoan and Keresen indigenous histories claim origins in the north.[30] This direction is opposite the southwestern location of the Hohokam.

28 <answersingenesis.org/go/theyhadnames/>

29 Ferguson, T.J., "Zuni Traditional History and Cultural Geography," in: Gregory and Wilcox 2007.

30 Ellis, F.H., "Laguna Pueblo," p. 439-441, in: Sturtevant 1979; Ortman 2012, p. 187-188.

Nearly surrounding all three of these Pueblo groups (Zuni, Keresan, Tanoan) at Contact were the Athabaskans (**Figure 70**). In **chapter 5**, we traced their origins to Canada. By the early A.D. 1200s, one of their relatives did battle with the Algic on the Plains.

When did the first Athabaskans reach the southwest? In time to scatter the local populations? The archaeology of the Mesa Verde region suggests an answer. In the late A.D. 1200s, the cliff dwellings emptied, never to be lived in again.

On the Cliff Palace tour, the park ranger chalked up the abandonment to drought. However, while lack of rainfall is the dominant explanation, it isn't the only one. The exact cause-effect relationship has been debated, even in mainstream literature:

> Historic records certainly indicate that droughts could have a devastating impact, but we also know that ancient Southwesterners suffered through many droughts without wholesale abandonments. Although AD 1300 is often cited as the date for the start of the abandonments, tree-ring dating consistently shows that construction activity diminished rapidly in the 1270s. The timing of the 'Great Drought' is such that it is more accurate to view it as the proverbial 'straw that broke the camel's back' rather than a calamity that initiated the abandonment process.[31]

What might the other straws have been? Recall the Algic-Athabaskan battle in the early A.D. 1200s (see **chapter 5**). Could these early Athabaskans have continued marching south so that, by the late A.D. 1200s, they were harassing the cliff dwellers? I don't have a hard time imagining such a scenario.

Up on the Plains, the opponents of the Athabaskans, the Algic, fought battles in the late A.D. 1200s *after they continued migrating.* Why not the *North Walkers* as well, *as they continued migrating to the south*? In fact, if they wanted to drive the cliff dwellers away, I don't think they would even need to descend the mesas into the dwellings themselves and fight on the cliff face. They would simply need to burn their crops on the mesa tops, starve them, and force them out. And if

31 Plog 2008, p. 152-153.

a drought happened to follow the raids, all the more effective for the Athabaskan purposes.

Curiously, right after the cliff dwellings around Mesa Verde were abandoned, the Pueblo population increased farther to the south. Mapping of the rises and falls of southwestern populations (**Figure 91**) suggests a cause-effect relationship.

In short, it seems like the Athabaskans arrived in the late A.D. 1200s. The Keresan and Tanoan look like they descend from those who fled the Athabaskans. Neither seem to have built the Hohokam culture.

At Contact, Hopis formed the final link in the circle around the Zuni, Keresan, and Tanoans. Earlier in this chapter we linked the Hopi mesas to the time of defensive positioning. They seem to be the vanquished, not the vanquishers. Hopis were not the Hohokam.

Going farther afield geographically, to the east of the Athabaskans at Contact were the Comanche (**Figure 70**). Their own histories put their origins near the Rockies, with a late A.D. 1600s/early A.D. 1700s-era arrival on the Plains.[32]

North of the Athabaskans were the Utes; northwest were the Paiutes (**Figure 70**). Both belong to the same subbranch of the Uto-Aztecan language family as the Comanche and Hopi. I doubt that Hohokam culture was built by any of these Uto-Aztecans.

To the south of the Athabaskans were the Pima and other Uto-Aztecans (**Figure 70**). Again, as per earlier discussion in this chapter, they seem to be conquerors, not the builders, of the Hohokam.

Just to the north and west of the former Hohokam areas, members of the Cochimí-Yuman language family resided (**Figure 70**). Archaeological work in this region is still early. But, apparently, the archaeology of the Cochimí-Yuman area bears some resemblance to pre-Classic Hohokam archaeology.[33]

That's a start.

32 Clark 1885, p. 118-119.
33 Shackley 2019.

In 1988, linguist Terrance Kaufman proposed a wider language grouping with relevance to this question. His inclusive group was labeled the *Hokan*. Kaufman linked Cochimí-Yuman with (1) a number of California language families, (2) several northern Mexican language families or isolates, and (3) one Central American language family (**Figure 92**).

Are there any historical events which might explain this distribution? Anything happening in Mesoamerica around the same time that the platform mound-building Hohokam arose? In short, *yes*.

When the Hohokam began platform mound construction in the southwest, in Mesoamerica one of the biggest empires, the *Toltecs*, fell.[34] Despite their collapse, the Toltecs would leave a permanent mark on Mesoamerican history:

> In their own annals, written down in Spanish letters after the Conquest, the Mexican nobility and intelligentsia looked back in wonder to an almost semi-mythical time when the Toltecs ruled, a people whose very name means "the artificers." Of them it was said that "nothing was too difficult for them, no place with which they dealt was too distant." From their capital, Tollan (Tula), they had dominated much of northern and central Mexico in ancient times, as well as parts of the Guatemalan highlands and most of the Yucatan Peninsula. After their downfall, no Mexican or Maya dynasty worth its salt failed to claim descent from these wonderful people.[35]

Who were the conquerors of the Toltecs? Currently, the answer is unknown. But the nomadic peoples of northwest Mexico have been suggested.[36]

What if the conquerors of the Toltec were the Hokan? Perhaps a western Mexican group split, with some of the party heading south and east to overwhelm the Toltecs, with another branch of the group

34 Aztec indigenous history suggests that the Toltecs fell in the late A.D. 1000s (Bierhorst 1992, p. 42). Archaeology suggests a date almost a century later, around A.D. 1150 (Coe and Koontz 2013, p. 170-178; Healan, Cobean, and Bowsher 2021). Either way, the dates are curiously close to the platform mound changes in the southwest.

35 Coe and Koontz 2013, p. 160.

36 Ibid, p. 178.

heading north into Arizona to start the construction of platform mounds.

All of this is speculative. But among the available options, the Hokan hypothesis is the most compelling.[37] I suspect that these were the cause of so much defensive angst in the rest of the southwest.

◂◂ ▪ ▸▸

*Terror* wasn't the last word for the southwesterners seeking to escape the Hokan. It may have been the most haunting, but I suspect it was not the most memorable. I tasted the latter early in the day at Mesa Verde. Before my tour at Cliff Palace, I set out for *Spruce Tree House* (**Figure 93**), one of the cliff dwellings for which reservations were not required.

Nestled under the overhang of a natural curve that connected one mesa peninsula to another, Spruce Tree House sat in the shadows, facing west in the late morning. Looking down on the site from the mesa top, I again kept my distance from the cliff edge. Hiking down the path to Spruce Canyon Trail, hoping to walk among the ruins, I found the distance between cliffside dwelling and canyon bottom to be much more manageable.

I reached a fork in the path and found out that the path to the ruins was closed. Disappointed, I set out on the trail instead.

On the path out, the distance between trail and canyon bottom increased. Plenty of blue sky swept away to my right. But the slope downward was always manageable, never a sheer drop. Even when the angle was sharp, plenty of vegetation reassured me that a misstep or a trip on the trail wouldn't mean certain death.

The views were indescribable. The atmosphere, delightful. Sunny with temperatures in the 80s, I relished picking my way along the rocky path. Nothing in the east compared to what I witnessed.

Alcoves were visible at frequent intervals. Some had the remains of brick-and-mortar constructions. At other places, I saw the black stains from ancient fires snaking up the face of the overhang or cliff. I

37 About a decade after Kaufman announced his Hokan grouping, in his book *American Indian Languages* linguist Lyle Campbell (1997), p. 296, wrote sympathetically about Kaufman's postulate.

was walking in the footsteps—perhaps quite literally—of the ancient Native Americans.

I think I could have handled raising a family at Spruce Tree House. The danger to my kids in their newborn and toddler years would have been manageable. Elementary years would have been filled with unequalled outdoor experiences. On the trail, I reflected on how fantastic it would have been to wake up each day and view the canyon vistas that I was taking in right then. Even after being forced to leave by a new threat from the north, I don't see how we'd ever lose the memories of our time in one of the most picturesque locations in all of North America.

I won't. I spent only a week in the southwest, and I already miss it.

# 10

## Solving the Mystery

April 22, 2024, I parked my 17-year-old blue Toyota Matrix on a shady single-lane dilapidated blacktop path that meanders through a grassy, lightly wooded 55-acre plot in the middle of my boyhood hometown. Since leaving Racine, Wisconsin, in 2003, I had moved three times. Crisscrossing the eastern United States and traversing over three thousand miles in the process, I once had called Massachusetts, and then Texas, and now Kentucky home. Today, I returned to Racine—on a mission.

I exited my vehicle near the northwestern corner of the almost trapezoid-shaped field. City streets paralleled each of the five sides, with older residences, businesses, and a 9-hole golf course across the way. One of the three major high schools was just a couple blocks down the road. It was around 3:00 p.m.

Outside the fenced-in property, students and pedestrians milled about. The sun was bright, ducking occasionally in and out of clouds. Temperatures were in the 60s, but the wind was crisp in the shade. I could hear the noise from traffic and the occasional siren of an emergency vehicle. But inside the fence, it was quiet. Hushed. I was in a cemetery.

I had been here once before. A friend of the family had died almost 30 years prior. I can still picture in my mind the wailing at her burial. Today, I found her crypt with little effort. It was marked with her name, along with the years of her birth and death. I did the math and realized she died when she was 43—the age that I am. But her grave was not the one I was seeking.

The crypt complex was only a small part of the sprawling city cemetery. Some of Racine's prominent businessmen—J.I. Case, of farm machinery manufacturing fame, and S.C. Johnson, notable for Johnson Wax—had their own mausoleums. They looked like small temples. Other grave markers spanned the whole gamut of designs. I saw everything from towering spires to simple headstones to flat foot-long stones flush with the ground. One high tower was crowned by a delicately carved, sad-faced girl scattering flowers. At another grave, a girl with downcast head rested her arm on a tall, rectangular headstone.

Names and descriptors recalled the pathos of the surviving family members. "Mother" and "father" were common. Simple words yet, in this context, pregnant with emotion. Another spire recorded the lifespan of the deceased down to the exact number of months and days. To confirm my suspicions, I scanned the fading writing for the age in years. Yes, it was a young boy in his teens.

I also noticed numerous graves with death dates in the early 1900s and even the late 1800s. Old graves. I guess I shouldn't have been surprised. The sign said the cemetery was established June 3, 1852. The lady in the office said that people had been buried here even before that date. Another sign mentioned graves of Revolutionary War veterans. I found a marker for one born 1761 who passed in 1836.

But I wasn't there today to look for the marked graves. I was after the unmarked ones—the ancient ones.

Growing up, I had no idea that the backyard of my sleepy town held the remnants of peoples who were contemporaries in time with the ancient Romans. *Mound Cemetery*[1] sits along Kinzie Avenue, one of the busier streets of Racine. My family and I had lived just over a mile away. We had driven down Kinzie uncountable times growing up, going to church in downtown Racine right along Lake Michigan. Looking for the ruins from peoples of millennia prior had not crossed my mind.

1 https://www.cityofracine.org/Cemetery/

You'd think the name of the cemetery would have provoked our interest. But none of us had given the mounds a passing thought.

Now our attitudes were different. One week prior to my arrival, my mother had beat me to the cemetery. She was so excited about the ancient graves that she wanted to see them as soon as she could—even though she had lived in Racine for decades.

What had changed? Two weeks prior to my visit, I had been up in Racine as a stopover on my way to Effigy Mounds National Monument. Once I was back home in Kentucky, I started digging for more information. I wanted to know the distribution of effigy mounds in Wisconsin. I came across an 1855 survey by Increase A. Lapham. His *Antiquities of Wisconsin* contained numerous plates with maps of mounds. Lo and behold, plate II was titled "Ancient Works Near Racine"!

*Were the Racine mounds still standing?* I had wondered, before my return trip to Wisconsin. I had known that many mounds identified in the 1800s had been plowed under. I didn't know if the mounds of my hometown had succumbed to the same fate.

Prior to my visit on April 22, I had known about a new technology that had been used to identify Indian ruins. *Light Detection and Ranging*, or *LiDAR*, involves shooting laser pulses from the air at the ground. After a lot of computer-based transformation, LiDAR produces exquisitely detailed topographic maps. I had already read reports of the successful use of the technology in Central and South American jungles. There, to the naked eye, dense vegetation had swallowed up and made invisible numerous ruins of the Mayans and of other civilizations. After LiDAR, the invisible became visible again. So I started asking if Wisconsin might have LiDAR-based maps.

It does. It turned out that the entire state had been mapped. I found and downloaded the nearly half a gigabyte-sized tiff image file for Racine County. There they were—about 11 pimple-shaped bumps in Mound Cemetery. I had shared these maps with my mother prior to my return trip. Next chance she got, she jumped in her car and checked them out.

On April 22, I made an initial loop around the cemetery, stopping at the 11 spots where the LiDAR map suggested an Indian burial. At the cemetery office, the worker showed me their property map highlighting 14 mounds. I guess I had missed a few in my initial search. She also told me one of the mounds I had visited was fake. The cemetery had constructed it at the northwest corner to draw attention to the site's history.

I went back to the mounds and saw two that I had overlooked earlier. One was no more than a foot or two tall and had a tree growing in the middle of it. Another was around the same height, which was generally shorter than the ones LiDAR had made obvious. In person, these larger mounds were three to five feet tall. Much easier to spot. The property map designated two more that I never could locate.

Lapham's 1855 map showed almost 100 mounds where Mound Cemetery and the surrounding area now sits. I guess most of them had been destroyed in the century and a half since his observations.

All of the mounds that I saw were simple, rounded, grass-covered hills. Some had large trees growing on them. None of the mounds were marked with gravestones. Except one. One of the taller mounds held an about 6-foot tall, grey, rough-hewn pillar with the words *Indian Mounds* etched on two sides. It sat in the shade of several trees, some of which had grey-ish colored trunks. It blended in so well that my mother had missed it on her visit.

The Indian mounds sat in sharp contrast to the more than 10,000 carved stone grave markers in Mound Cemetery. No dates of births and deaths. No doleful phrases to mourn the dead.

Visually, the mounds were unimpressive. They don't strike the eye as unusual feats of construction. I learned that the hard way when I was fooled by the artificial mound.

So what was the draw? For both my mother and me, it was the discovery, prior to our trips, of what you've read about in preceding chapters: A *new history* of pre-Columbian North America, a new framework that represented a sharp departure from our knowledge several decades prior.

The new framework gave context, narrative, and meaning to a site that, in our minds, had little before.

◀◀ ▪ ▶▶

A second draw to Mound Cemetery in Racine was the mystery. The estimated date for the Mound Cemetery structures is before the arrival of haplogroup Q. *In other words, my hometown mounds were built by people I know the least about.*

In the 1800s, one of Increase Lapham's colleagues excavated one of the biggest mounds. Inside were "the skeletons of seven persons, buried in a sitting posture, and facing the east."[2] Why? Despite the progress described in previous chapters, I still don't have the answer.

But I'm getting closer.

Several months prior, on Wednesday, December 13, 2023, I found myself sitting among a crowd of people in a Holiday Inn ballroom in Rapid City, South Dakota, as part of a three-day-long Native American gathering. I was on a research trip. I wanted to know the origins of the early Native Americans—the people of the Hopewell and Poverty Point cultures (see **chapter 2**). To date, the answer had remained elusive. Their genetic echo was still missing. But I thought this trip might put me on my way to uncovering the answer. The Lakota attendees at this meeting held one of the keys.

Four years prior, I would have never dreamed that I'd be a guest of the Lakota. In the fall of 2019, when the initial Y chromosome DNA results for Native Americans were first emerging, I enthusiastically started calling Native American reservations cold turkey, with hopes of announcing the new research findings to them. At least one hung up on me.

A couple of years later, I released a video on YouTube on genetics and Native American history. I had just written a book on the human genetics of world history, with one of the chapters covering pre-Columbian American history. The video summarized the chapter and invited Native Americans who wanted to participate in future research to contact me. I didn't know if any would.

Within the first two months following the release of the video, about six Native Americans did. One freely announced to me that I had adopted the appropriate strategy for engaging them. "We contact

2 Lapham 1855, p. 7.

you; you don't contact us. Otherwise, we won't speak to you" was the gist of her exhortation. I received it.

About a year later, I started releasing more videos, detailing some of the early results that this book describes. Each time, I invited Native responses. I now have contacts in more than 35 Native nations. One of them—the Lakota—invited me to their December 2023 meeting.

In discussions with the Lakota, I learned that they have astronomical links back to the B.C. era.[3] They describe the arrangement of certain constellations in a manner that makes sense only if we dial back the history of the sky a couple of millennia. In other words, the Lakota, perhaps more than any other living North American Native American group, might possess the long-lost DNA links to the earliest Native Americans.

◂◂ ▪ ▸▸

In previous chapters, I've said that we possess no Y chromosome lineages older than haplogroup Q. A more precise way to say this would be: We possess no Y chromosome lineages older than haplogroup *Q at the moment*. It's possible that lineages from the earliest Native American peoples still exist but haven't been found yet. It's possible due to a quirk of statistics.

As an illustration, let's consider the recent population history of one of the most numerous surviving Native peoples, the Navajo. In 1868, the Navajo population numbered 9,000.[4] By the year 2000, the number had ballooned to more than 175,000—a nearly 20-fold increase.[5] The 2010 U.S. census[6] suggests that the number might now exceed 300,000.

For starters, let's assume that the Navajo population has always been 50% males, 50% females. Therefore, the number of living Navajo men today is half of the number implied by the 2010 U.S. census (i.e.,

3 Goodman 2017, see especially his Appendix B.
4 Obtained from http://www.navajobusiness.com/pdf/FstFctspdf/Tbl3GrwthRate.pdf, accessed 11/25/19.
5 Ibid.
6 2010 Census of Population and Housing, American Indian and Alaska Native, Summary File: Technical Documentation, U.S. Census Bureau, 2012. On October 10, 2019, I downloaded Table 1. American Indian and Alaska Native Population by Tribe for the United States: 2010 (Internet release date: December 2013).

150,000). In 1868, half of the 9,000 people (4500) would also have been males.

This history of population recovery statistically constrains the branches of the Y chromosome family tree. Each individual male alive today would belong to a branch on the Y chromosome family tree. In theory, each male would have his own tree tip. Practically, this means that, today, there are going to be about 150,000 Navajo Y chromosome branches in existence. In 1868, there would have been 33 times fewer—just 4,500. Or, in terms of percentages, only 3% of today's branches would have existed just 150 years ago.

How can this be possible? In math terms we can restate the question: How do we reduce the number of today's branches on the Y chromosome tree by 97% in 1868? The answer is elegant yet simple.

Consider the following thought experiment: Try counting all of your biological ancestors. Begin with your parents and go backward in time. You have two parents. Your parents each had two parents, which means you had four grandparents. Your four grandparents each had two parents, leaving you with eight great-grandparents.

Do you see the pattern? The number of your ancestors doubles each generation (**Figure 94**).

Now take the thought experiment back about 900 years. To keep the math simple, let's say all of your ancestors had their first child by age 25. Each century, then, would see four generations. Travel back nine centuries, and you go back 36 generations. By this point, your ancestors would number a whopping 68 *billion*.

If this seems impossible, it is. The world population today is just past 8 billion. Nine hundred years ago, the population would have been much less.

Clearly, something has to change. But we can't change the raw numbers. They flow from basic facts of biology. Humans don't reproduce asexually but sexually. Doubling your ancestors every generation is a requirement.

Stumped? Think through the solution in terms of branches on your family tree. One generation ago, your family tree had two branches—one from mom, one from dad. Two generations ago, there were four—one for each of your four grandparents. At 36 generations ago, there were more than 68 billion. The solution is one that reduces this number.

We can't reduce this number by removing or killing off branches. The nature of human reproduction forbids this. Each individual in each generation must have two parents.

Instead, to reduce the number of branches, *we'll connect them* (**Figure 95**). In other words, your father's side of the family tree and your mother's side of the family tree *must connect in the recent past*. The vast majority of today's Navajo Y chromosome branches join in less than two centuries.[7] The math requires it.

Now let's explore how this percentage affects our ability to see history in Navajo DNA. Let's say that we sample the Y chromosomes of 100 Navajo men today. By today's standards, this is a respectable study size. These 100 men would be represented by 100 branch tips on the Y chromosome tree.

But what if we were to go backward in time along the tree? How would the branches change? Given the history of the Navajo population, we would find that 97 of the branches (i.e., 97% of 100 men) would connect by 1868. In other words, going back in time just 150 years, we would have only three branches remaining. These three branches would be the only genetic tools that we would have to explore the past deeper than 1868.

Do you see the dilemma? The statistics of the Native American population recovery make it hard to dig into the ancient past. You really need to test the DNA of many more men to have any statistical hope of digging into ancient American history.[8]

How many men do we need? Let's run the math again with bigger starting numbers. Let's say that we sample the Y chromosomes of, not

7 This isn't a phenomenon unique to the Navajos or to Native Americans. Within the most recent 600 years, the *global* population has increased from an estimated 350 million people in A.D. 1400 to over 8 billion today (McEvedy and Jones 1978.) This is more than a 20-fold increase. Consequently, looking backward in time from the present, 95% or more of today's branches on the global family tree connect by A.D. 1400.

8 Again, this same dilemma faces investigators seeking to recover the ancient human history of the globe. Because the global population has increased 20-fold in the last 600 years, 95% of the branches on the global family tree represent the most recent six centuries.

100, but 100,000 Navajo men today—a huge study. Given the history of the Navajo population, we would find that 97,000 of the branches (i.e., 97% of 100,000 men) would connect by 1868. Going back in time just 150 years, we would have only 3% of the branches remaining. But this time, 3% of 100,000 represents, not three total branches, but 3,000. These branches give us far more tools with which to explore the past deeper than 1868.

Currently, only a small percentage of Native Americans have taken a Y chromosome test. Deeper sampling of these men might yield a breakthrough. In other words, the genetic echo of ancient North American peoples might still exist among living Native North Americans.

And it could be among any of them. In **chapter 2**, I focused on the archaeology of just two ancient North American cultures—the Hopewell and the Poverty Point culture. Both were located around or east of the Mississippi, but evidence for an ancient human presence exists all over North America.[9] Even in the southwest, hunters and gatherers roamed during the B.C. era. Surviving lineages might be found among any of the surviving peoples depicted in the **online Contact-era map**.

But for research, invitations matter, and my main invitation came from the Lakota.

◂◂ ▪ ▸▸

At the December 2023 meeting, I presented a brief, 30-minute set of tips for Native Americans navigating the field of DNA testing (see a version of this in **Appendix C**). My talk was followed by open mic Q&A. The one question I received surprised me.

My past experience as a creationist led me to expect a question about evolution. But no one did. Instead, one gentleman asked if what I taught contradicted what his family had told him about their origins.

I learned that Native peoples don't use evolution to explain their origins. Because they hold to the existence of a Great Spirit, you could describe their worldview as creationist in nature. They also believe that they literally came from the earth and are very opposed to any talk of

9 e.g., see Anderson, D.G., "Pleistocene Settlement in the East," Figure 9.1, p. 97, in: Pauketat 2012.

Native origins via a migration across the Bering Strait. This is what the gentleman was referring to when he asked about contradictions.

I don't think I gave a good answer. I wasn't prepared for this perspective on the past. Prior to the Lakota trip, I had visited Latin America. By appearances and DNA, Latin Americans retain a strong indigenous connection. However, in terms of culture, Latin American universities teach Marxism, Darwinism, and other strongly European doctrines. In terms of worldview, Latin America as a whole is pervasively Western. Being among the Lakota was my first experience of a thoroughly indigenous worldview.

Prior to the conference, I had tasted indigenous outlooks indirectly from the news or from books. I was familiar with the idea that some Native Americans insisted on always having been in America. Being ignorant, I guess I had not taken the sentiment seriously. I suppose it sounded too politically convenient. But I was wrong.

Let's state the dilemma for Native Americans as bluntly as possible. This book describes a synthesis of genetics, archaeology, linguistics, and indigenous histories. The latter might be attractive to Native peoples. But the genetics are thoroughly grounded in a young-earth creationist framework.

What framework? Again, let's be as blunt as possible. My calculations and conclusions explicitly invoke a timeline for human civilization that is only 4,500 years long. This is radical, but I've laid out numerous, independent scientific reasons why this is true.[10] The Y chromosome tree I'm using finds an explicit echo in the genealogies of Genesis 10. The plain ramifications of my genetic framework are that every people group goes back to the Middle East, to Noah's family, and does so within just a few thousand years. Surely the Bering Strait will have to be crossed at some point (see also **Appendix D**).

Prior to the conference, I had read about the struggle of a Mescalero Apache elder named *Bernard Second*. He had wrestled, not with creationist ideas, but mainstream evolutionary ones. The mainstream authors, who spent 20 years with him, described a sequence of events that took on new relevance after my time with the Lakota:

10 Jeanson 2022; see also **Appendix A**.

> Bernard had long been aware of the standard anthropological view that the ancestors of Native Americans were ultimately derived from Asian populations who had migrated to the New World via the Bering land bridge. Nevertheless, while he respected our views about Native American origins as the product of serious professional study, he remained skeptical. After all, he knew that his own people had originated 'here,' on this continent, on the shores of a Lake You Can't See Over, in a Land of Ever Winter.[11]

Yet Bernard eventually changed his view, for reasons that surprised me. Bernard was brilliant, speaking more than five languages, including Japanese. Bernard's skill in the latter ended up playing an instrumental role in changing his mind about his people's origins:

> One summer afternoon...he [Bernard] had received a group of visitors...including a man from Japan. ... [Bernard] relished the opportunity to practice the language and learn about Japanese culture. He spent a couple of hours in deep conversation with the Japanese man. ...
>
> Clearly, there were some similarities in physical appearance between the Japanese man and at least some Apache people. But Bernard was more impressed by some specific similarities between their two cultures. The visitor traced his family roots to the samurai tradition, and even though the samurai influence in Japanese society declined in the late nineteenth century and officially ended after World War II, the family maintained knowledge of its ancestral traditions. Bernard and the Japanese visitor spoke at length about what it means (or what it meant) to be a warrior in the traditional sense. To Bernard's surprise, there were a number of detailed similarities between the samurai code of *Bushido* and the Mescalero Apache 'Way of the Warrior.' The similarities included not only the sorts of behaviors expected of warriors in each society (such as the importance of duty, honor, and self-sacrifice) but also the underlying spiritual insights on which the codes are based. ...

11 Carmichael, David L. and Farrer, Claire R. "We do not forget; we remember: Mescalero Apache origins and migration as reflected in place names," p. 195, in: Seymour 2012.

> Until meeting the Japanese visitor, he had had no firsthand exposure to any Asian culture that might lend credence to the notion of an ancient connection between his people and an Asian population. But after that meeting, the possibility of a connection to another continent became an observable reality. Moreover, it became a matter of personal intellectual interest for Bernard to reconcile the existence of such a connection with his knowledge of the origin of his people. He accomplished the reconciliation in a very simple yet elegant way—by drawing a distinction between biological heredity and cultural identity. That is, "We became a people near Lake Athabasca, but we have earlier biological connections to Asian populations."[12]

I surmised that my presentation to the Lakota in 2023 was the first time they had heard views like mine. It wasn't just because the research itself was new. It was also because evolutionary ideas, not creationist ones, currently dominate Western culture. Surely, they would need time to process.

◂◂ ▪ ▸▸

A few weeks after the December 2023 meeting, I spoke with the executive director again over the phone. He communicated that many Lakota still have deep reservations about DNA testing. I told him that I thought this was completely understandable and that each person should do only what they're comfortable with.

My conversation with the Lakota remains ongoing. I don't have any timeline for its completion. I recognize that establishing trust takes time. They're getting to know me, and I'm getting to know them. It's unpredictable, like all relationships are.

Currently, I am actively exploring research options in Latin America. Given the history of connections between peoples north and south of the Rio Grande, it's just as plausible that the most ancient, elusive Y chromosome branches still exist in Latin America as much as they do in North America.

---

12 Carmichael, David L. and Farrer, Claire R. "We do not forget; we remember: Mescalero Apache origins and migration as reflected in place names," p. 194-195, in: Seymour 2012.

In addition, the results in this book raise a whole new set of questions, for which Latin America holds the key. Which Mexican populations were the source for the Natchez? For the Muskogeans? For the Hokan and other peoples of the southwest with plausible Central American links? What is the history of these peoples before their migrations north? The genetics of Latin America promise a wealth of new insights.

And whole new chapters on the ever-growing pre-European history of the continent that I call home.

# Afterword: "Kill the Indian, Save the Man?"

I want to see Native Americans become Christians. Granted, as a Christian, I want to see people of every ethnicity be saved. But in a book like this, I feel my desires renewed specifically for the Native American community.

Perhaps this offends you. Yes, I've made a rather blunt statement. But if you're Native American, my statement might also leave a bitter taste in your mouth. "Oh, here we go again. He's going to try to strip me of my Native identity—another form of the old practice of 'kill the Indian, save the man.'"

I'd like to offer two responses. First, I'd like to justify my bluntness. It flows from my core convictions. I believe that Christianity is true. I believe God created a perfect world about 6,000 years ago. I believe He created two sinless people in the beginning—Adam and Eve—and forbade them from disobeying Him, under the threat of death.

This prohibition was anything but harsh in light of God's nature. The entire Bible sings the glory of God—an indescribable, overwhelming, magnificent, pure perfection beyond human imagination. Disobeying His commands is an indescribable evil.

I believe Adam and Eve did just that, bringing judgment on themselves. The Bible holds Adam especially responsible: *"through one man sin entered the world, and death through sin, and thus death spread to all men, because all sinned"* (Romans 5:12; NKJV). I believe that Adam's sin brought judgement on all of humanity. We have continued their sinful practices, and we await God's judgment: *"for all have sinned and fall short of the glory of God ... the wages of sin is death"* (Romans 3:23, 6:23; NKJV). I believe this is true for every person, every nation, every tribe that has ever walked the face of the earth. A fiery hell—appropriate punishment for those who desecrate God's glory—awaits everyone.

I also believe that God made a way of escape. He sent His Son, Jesus, to take the punishment for us. The Bible uses language that explicitly connects Jesus back to Adam: *"If by the one man's offense death reigned through the one, much more those who receive abundance of grace and of the gift of righteousness will reign in life through the One, Jesus Christ"* (Romans 5:17; NKJV). *"And so it is written, 'The first man Adam became a living being.' The last Adam became a life-giving spirit"* (1 Corinthians 15:45).

Those who repent of their sins, and believe that He died on the cross and rose again, escape God's wrath and will one day enjoy Paradise—the New Heavens and the New Earth.

Therefore, I want to see Native Americans (as well as all nations and ethnicities) become Christians. I don't think you could fault me for this.

"But what about Native culture? Don't you want to take that away?" I gained a new perspective on this question after attending the Lakota Treaty Council conference in December 2023.

Growing up in a majority culture, I've never had to wonder about loss of my own culture. It's never been threatened. But being physically present at the Lakota conference, I glimpsed the tension inherent in being a minority culture.

From the podium, the older generation of Lakota lamented how difficult it was to pass on Lakota culture and language to the younger generation. With little reflection, I thought of several reasons why. Lakota youth live in a nation with English as the official language. Local and national news is dominated by Western issues with Western people. Entertainment features Caucasians, some African-Americans, and some Latinos. Native North Americans have almost no visual signature in Western culture. Yet the peer pressure—implicit and explicit—to embrace it is relentless.

The conference sought to maintain elements of Lakota culture. It opened with a Lakota sacred ceremony. Each day featured songs around a drum circle. Parts of the conference were conducted in the

Lakota language. In terms of identification, I was one of the few attendees without a descriptive, Native-style name.

Despite these efforts, it was impossible to miss the non-Lakota influences that were creeping in. We were conducting the meeting in a Western hotel with Western comforts. We ate Western food and conversed primarily in a Western language. Dress was a mix of Western and Native styles.

After the conference, the cultural struggle hit me in a new way. I had a hunch regarding Native population numbers. I dug a little deeper to grasp the precise figures. The 2010 U.S. census[1] puts the combined "American Indian and Alaska Native" population at just 0.9%. African-Americans are 14 times more numerous. Yet I grew up associating the word "minority" with African-Americans, not Native Americans.

Some African-Americans had a whole summer of riots in 2020, prompted in part by the death of George Floyd and in part by lingering bitterness from racism and the practice of slavery. Native Americans, despite being a minority in a much more severe mathematical sense, have had no such attention. Surely Native Americans notice this discrepancy. Surely it fuels fears that Native culture will one day disappear.

◂◂ ▪ ▸▸

Does the Bible have anything to say about this fear? Does it suggest that non-European cultures should reject their own cultures when becoming Christians? Does the Bible teach, "Kill the Indian, save the man"? Does it offer any hope to minority cultures? I eventually found the answers, but the implications caught me by surprise.

The last Book of the Bible, Revelation, reveals a future scene with direct relevance to our discussion. The writer, John, saw the future New Jerusalem:

> *The city had no need of the sun or of the moon to shine in it, for the glory of God illuminated it. The Lamb is its light. And the nations of those who are saved shall walk in its light, and*

1 Humes, K.R., Jones, N.A., and Ramirez, R.R., issued March 2011, "Overview of Race and Hispanic Origin: 2010," 2010 Census Briefs, U.S. Department of Commerce, Economics and Statistics Administration, U.S. CENSUS BUREAU, Table 1, p. 4, https://www.census.gov/content/dam/Census/library/publications/2011/dec/c2010br-02.pdf, accessed July 10, 2024.

> **the kings of the earth bring their glory and honor into it.** *Its gates shall not be shut at all by day (there shall be no night there).* **And they shall bring the glory and the honor of the nations into it**[2] [emphasis added].

I wondered: What does this mean? Will the glory of ancient Rome be present? The glory of ancient Greece? Of the Han Dynasty in China? Of the Guptas in India? Of the pre-European West African kingdoms? Of the Lakota? Or will the glory be strictly whatever survives to the end of time? I had a hard time believing the answer was the latter. Surely the glory of peoples who have come and gone, and of those alive right now, will be present.

Would every single aspect of every single culture be preserved? Take the glory of the Roman empire as an example. Would the "glory" include their practice of slavery? Would the New Jerusalem celebrate this barbarity? What about cultures that practiced head-hunting? Will the people of the New Jerusalem engage in this pursuit? What about cannibalism? What about Nazi Germany? Would the Holocaust continue in the New Jerusalem? The rest of the Bible would answer these questions with an unambiguous *no*. The Book of Revelation implies the same. Immediately following the quoted section above, the verse reads:

> *But there shall by no means enter it anything that defiles, or causes an abomination or a lie, but only those who are written in the Lamb's Book of Life.*[3]

In other words, Christianity doesn't teach "Kill the Indian, save the man." Christianity rebukes it. Because *God saves the Indian and the man.*

◂◂ ▪ ▸▸

I seek to align my desires with God's. If He wants to save the Indian and the man, so do I. One of the purposes of this book is to provoke the recovery and dissemination of pre-Contact history. I want to see it respected and preserved.

---

2 Rev.21:23-26; NKJV.
3 Rev.21:27; NKJV.

I also want to see the Lakota language maintained, along with indigenous history. Language contains critical clues to history. Indigenous accounts of ancestral migrations do, too. Any chance that I have to engage elders from an indigenous group, I plead with them to write down their history, to prevent its permanent loss. In previous chapters, you've seen how vital a role these histories played. I don't want to see this rich resource disappear.

I've also learned to appreciate elements of indigenous culture that aren't present in the West. For example, when I was at the Lakota conference, I was struck by a specific Lakota practice. It seemed that Lakota under age 50 who rose to speak began by apologizing—apologizing to their elders for standing in front of them and speaking. I don't see any such respect for the aged in Western culture. It seems we Westerners could learn a lesson or two from the Lakota.

None of these desires to preserve indigenous history, culture, and language remove my desire to see Native Americans saved from their sins. God desires it. So do I.

I hope you'll consider what the Bible says about your state before God. All of us have offended Him. All of us justly deserve His wrath. But God made a way of escape through His Son, Jesus. Those who believe He died and rose again, who repent of their sins, stand righteous before God.

I've embraced this. I hope you will, too.

# Appendix A: "How Do I Know This Is True?"

I have written this Appendix for a slightly different audience than the audience for the main text. I presume that readers of this Appendix desire a more detailed justification for my conclusions. Consequently, I've added more scientific depth.

Overall, my goal in this section is to answer a single question: Why should a reader believe what I've concluded about the pre-Contact history of North America? My answer in summary is: Because my conclusions are based on more than a decade of a consistent pattern of making testable genetic predictions, and then seeing these predictions fulfilled by later experiments. These have led to more predictions and more fulfillments. I've also witnessed these genetic findings match data from other fields, whose own data harmonize across disciplines. In short, I think my conclusions are reliable *because they are working well.*

I've focused on this aspect of my work for specific historical and practical reasons. Critics of creation science have demanded exactly what I'm providing:

> *The most important feature of scientific hypotheses is that they are testable*[1] [emphasis theirs].

In other words, in scientific debate, evidence is important, yes. But evidence alone is insufficient. It must be continuously evaluated. Science—as a discipline—lives in the perpetual present. All claims are provisional. They are subject to on-going experimental evaluation. The best scientific claims are those which survive the most attempts to disprove them.

1 Futuyma and Kirkpatrick 2017, p. 578.

Why focus primarily on genetics? In terms of logical dependencies, my conclusions flow heavily from this field. My claims about multiple migrations to the Americas certainly do. Specifically, they rest on the dates for Native American branches in the Y chromosome DNA-based family tree. My claims about indigenous histories are also founded on these DNA findings. If the dates I've assigned for the Y chromosome branches are invalid, much of the narrative dissolves. Clearly, then, it's critical to evaluate how and why I derived my dates for the branches.

These dates themselves rest on two more foundational conclusions. They depend on the overall timescale for the Y chromosome DNA-based family tree. They also depend on the specific root (beginning) position in the tree. On both of these points, my conclusions differ strongly from mainstream ones. For our discussion, then, we need to know why I disagree. We also need to know whether my reasons are valid.

I'll let you decide the answer to the latter question. With regard to the former, what follows is a narrative of my multi-year experimental progress. It starts a bit afield from the Y chromosome, for reasons that will soon become clear. I want you, the reader, to see what questions I asked and where the results led me. Then we can return to the question of their validity.

To keep this Appendix from waxing excessively long, I've retold the narrative in a high-level way. For genetics specialists, I've provided copious footnotes to the technical publications. These contain the complex methodological details and formulas. They also contain the raw datasets for anyone wishing to repeat and cross-check my experiments.

## Deriving the framework for the Y chromosome tree

Fifteen years ago, I dipped into the field of human genetics with the main type of dataset available to researchers: The maternally inherited *mitochondrial DNA (mtDNA)*. Human genetics was dominated by mtDNA for practical reasons. It represents just a miniscule fraction of the total DNA of an individual—less than 0.001%. *Autosomal* DNA—the DNA that we inherit from both parents—represents about 99%. Y chromosome DNA constitutes the remainder.

Because mtDNA is so small, it is much cheaper and easier to sequence. At the time I started working on the genetics of human history, thousands of human mitochondrial DNA sequences were present in the publicly available DNA datasets.

With respect to the timeline of human origins, I looked for a specific type of published experimental result. Parent-offspring mutation rates are easy to obtain. They're also largely free of encumbering assumptions. If we could determine the rates at which mtDNA was changing now, perhaps we could extrapolate this rate into the past to determine the time of a species' origin.

In short, the mitochondrial data were clear: Human origins traced back only a few thousand years. This conclusion was consistent across other species, phyla, and even kingdoms of life. It also made testable predictions for the future. In addition, the logic I used to reach my conclusions had broader ramifications. It undercut the assumptions used in other fields of science to argue for an ancient origin. I was cautiously optimistic that I had stumbled upon something real.[2]

Eventually, I began to analyze these same two questions in another genetic context—that of autosomal DNA. Here, the data for the timescale contradicted the data for mitochondrial DNA. I evaluated several possible solutions.

The mainstream view faced the same problem that I did—contradictory conclusions from mitochondrial and autosomal DNA. The published mainstream attempts at reconciliation tended to rely on special pleading. They did not represent a coherent explanation. I also did not find testable predictions.

In contrast, the model I had been chasing so far provided an elegant solution. It was free of special pleading. It also made rigorous testable predictions. By objective standards, these data pointed towards a timescale of only a few thousand years for human origins. I published these results in 2016.[3]

Right around this time, large datasets of fully sequenced Y chromosomes began to be published. I and another colleague began to explore

2 Jeanson 2013; Jeanson 2015a; Jeanson 2015b; Jeanson 2016. See also chapter 7 of Jeanson 2017.

3 Jeanson and Lisle 2016. See also chapters 7-8 of Jeanson 2017.

questions of timescale anew, using both mtDNA and Y chromosome results side-by-side.

One of the first specific questions we sought to analyze revolved around the Trans-Atlantic slave trade. Precise datasets on population numbers existed. We could look up exactly how many Africans were captured, how many survived the Trans-Atlantic voyage, and how many disembarked at specific geographic locations. We could plot all of these numbers on a specific timeline.

In addition, we had access to genetic datasets of indigenous African and African-American individuals. Mitochondrial DNA and Y chromosome DNA data were available for both populations. We sought to answer a simple set of questions: When did the African-American and indigenous African lineages diverge? Was the genetic date consistent with the known historical data? We tested the genetic data under a whole slew of hypotheses.

To calculate dates of genetic divergence, we had to assume an overall timescale and root (beginning point) for the Y chromosome DNA-based family tree. We tested the mainstream timescale and the recent origin timescale. We also tested a variety of root positions. Objectively, the recent timescale predicted the data the best. We could not, however, firmly nail down the exact root position in the tree. Instead, we identified a range of positions that fit the data.

I was thrilled. Yes, the precise root of the tree remained an open question. But the question of overall timescale now had robust answers. We had three independent genetic compartments—mtDNA, autosomal DNA, and Y chromosome DNA—telling a consistent story.

We never published the results. As we were writing up the paper, I kept taking a critical eye to it. I wanted to anticipate the reviewers' objections in advance and make sure we had answers.

I couldn't answer one big objection. I imagined that the reviewers might be willing to concede that our data for the Trans-Atlantic slave trade were solid. But what about the implications for the rest of the tree? My results dated the beginning of the tree as recent. Surely this would radically alter the separation times for branches all throughout the tree.

At this stage, I narrowed my focus to the Y chromosome DNA-based tree. I had observed that the statistics for the mtDNA-based tree were inferior to the Y chromosome one. For example, consider the standard deviation for each tree. For the mtDNA-based tree, it was around 20% to 25%. For the Y chromosome DNA-based tree, it was around 4% to 6%. Clearly, the Y chromosome DNA-based tree was a far superior, far more precise tool. I could use it to derive much more precise branch dates.

## Native American genetics enter the picture

I also narrowed my focus to resolving one specific historical question. I had read about the post-Columbian Native American population collapse.[4] In addition, I knew that some groups showed significant post-collapse recovery. I now went looking for the genetic signals of these events.

Among the Native American Y chromosome branches, I discovered what I was looking for. The collapse and recovery exactly matched the historical data. But only if I did my calculations from the recent-human-origins model.

I didn't publish the findings right away. Instead, with newfound confidence that I was on the right track, I revisited the published literature on father-son Y chromosome mutation rates.

In short, I found that high quality[5] studies showed fast mutation rates. Low quality[6] studies showed slow mutation rates. The former was in line with the recent-origin-of-humanity model. To my shock, the mainstream community dealt with this contradiction of their model by filtering out the contrary data—removing results that didn't fit their preconceived expectations.[7] (I'm not exaggerating; see the paper in the footnote for the primary references.)

Father-son mutation rates represent just one small snapshot in time for the long history of humanity. I needed to know if this fast mutation rate had been consistent throughout human history. I tested this hypothesis by (1) reconstructing the history of population growth

---

4 Mann 2005.
5 In technical terms, *high coverage*.
6 In technical terms, *low coverage*.
7 Jeanson and Holland 2019.

from the Y chromosome tree, based on a recent origin, and then (2) comparing it to the history of population growth based on archaeology and historical records. The match was a tight fit.[8]

After these discoveries, I published my findings from the Native American Y chromosome branches.[9] I also returned to the question that had been nagging me: What to make of the new dates for all the branches in the global family tree? I had a hunch, based on the recent origin hypothesis.

The recent origin conclusions were consistent with a specific inference from Scripture. Genesis 6-9 claim that all modern humanity traces their ancestry back to eight people (Noah, Noah's wife, Noah's three sons [Shem, Ham, Japheth], and their wives) around 4,500 years ago. These same chapters claim that all pre-Noachian civilizations were destroyed in a global Flood. If true, then the history of civilization was only a few thousand years old. Naturally, then, I would expect all the new dates for the branches in the global family tree to reflect the history of civilization.

They did.

From regional population growth curves to the timing of population splits to the geography of modern branches, the patterns throughout the tree lined up with the known history of civilization. I published a book-length treatment of these results in 2022.[10]

The book also detailed the solution to another set of questions. Prior to 2021, I was aware of several specific elements of the biblical anthropology. I knew that Genesis 10 listed the male descendants of Noah's three sons, Shem, Ham, and Japheth. I knew that this genealogy made testable predictions. But I hadn't found the fulfillment of them.

You can see this in my publications from the pre-2021 period. I entertained multiple possibilities for the root of the Y chromosome tree. I also had no explicit statements on how, specifically, the details of the biblical narrative played out. No firm conclusions on Noah, Shem, Ham, and Japheth. No drawn-out inferences on the fates of the Patriarchs listed in Genesis 10.

8 Jeanson 2019.
9 Jeanson 2020.
10 Jeanson 2022.

Then in the spring of 2021, I began to draw out more specific implications of Genesis 10. I then compared my newly formulated hypotheses from the biblical data to the Y chromosome branches. At that point, I already had in hand a comprehensive set of explanations for the branches, based on matches between the history of civilization and the Y chromosome tree. The biblical hypotheses matched the Y chromosome data.

As a result, I was able to pinpoint the root of the Y chromosome tree. Prior to the biblical discoveries, I had a range of possible root positions but no definite conclusions. The biblical data, with its precise generation-by-generation data for the earliest periods of human history, resolved the question unambiguously.

I included all of these results in the 2022 book.[11]

## The critics

At this point, I think it's important to pause and reflect on the significance of this series of findings. Notice that it's not primarily a refutation of predictions from a hypothesis of an ancient human origin. I didn't spend 10+ years primarily picking apart the evolutionary model. Are there a multitude of results that challenge the mainstream view? Yes. Did I focus on this aspect of the work? No. Instead, I showed via one discovery after another that my own recent origin hypothesis was working. The challenge for my opponents is: *If my ideas are wrong, why do they keep working so well?*

If this strikes you as an unusual take on the creation-evolution debate, it is. For 40 years, much of the focus of the creation-evolution debate has been on identifying, and evaluating, problems with evolution. For forty years, evolutionists have insisted that this strategy was insufficient. It wasn't enough to hamstring evolution. Instead, in print and in the court system, evolutionists demanded that creationists publish and test predictions based on creationist views.[12]

Now that I have, and now that I've shown a consistent pattern of the success of these predictions, we find ourselves at a remarkable crossroads. How will evolutionists respond? Will they adopt the old

11 See especially chapter 13 of Jeanson 2022.

12 E.g., see: Eldredge 1982, p. 80, 138; *McLean v. Arkansas Board of Education*; Futuyma and Kirkpatrick 2017, p. 578, 583-584.

creationist strategies? Will they try to poke holes in my ideas rather than publish better testable predictions of their own? How will they explain the multi-year scientific successes of my model? Some of these questions remain outstanding; others have begun to receive answers.

To date, few mainstream scientists seem to be aware of my findings. The handful of professional biologists who have responded never bothered to engage my arguments. One rather vocal PhD stooped even lower, implying that I was wrong because I disagreed with the textbooks and with the mainstream literature—as if these two sources of knowledge were not to be questioned.[13] In short, there have been few mainstream attempts to rigorously test and contradict my claims.

Given the history of hostility toward creation science from the mainstream community, I'm somewhat surprised by this tepid response. I would think that the critics would be eager to take my published predictions, test them, and try to refute creation science once and for all.

For example, one of the biggest predictions relates to the rate at which we discover new Y chromosome branches. I've published a specific, mathematically precise formula to anticipate where, when, and how often in the Y chromosome tree new branches will appear.[14] Testing this hypothesis would be the most straightforward way to reject my claims.

I suspect, however, that more is afoot than simple ignorance. My predictive formula is now about five years old. I've seen it fulfilled in the academic literature.[15] I've also seen it play out in the FamilyTreeDNA Discover database—the biggest publicly available repository of Y chromosome results on earth.[16] I wonder how many critics are silent because my testable predictions are turning out to be true.

---

13 E.g., see my response here: https://www.youtube.com/watch?v=1sbiHj_0r4o&list=PL1v9pqs4w1mzxnSlpFs-QmE-cAFm12398&index=16

14 Jeanson 2019. For the "where" part, see Jeanson 2022.

15 E.g., see discussion in Jeanson 2022, especially Appendix B therein.

16 https://discover.familytreedna.com/

## Evolution

Up to this point, I haven't dealt explicitly with the question of evolution. You might have already wondered where the evidence for evolution fits in this larger discussion.

In brief, I meticulously explored this question in a separate book.[17] From biogeography, to homology, to "bad design," to transitional fossil forms, to more obscure evidence for evolution both from Darwin's day and the present, I walked through the data step by step. I won't try to cram the conclusions in these pages, other than to highlight a critical observation: For more than 150 years, evolution has marshalled evidence for itself by excluding creationist ideas from the outset. Evolution has *not* gained support by performing experiments that test these two ideas as competing hypotheses. Once we do, the entire dynamic of the debate changes.

## What about ancient DNA?

This book on Native American history, and the slew of papers and books that precede it (see previous subsections), are based on the *exclusion* of ancient DNA from the analyses. For example, I don't include Neanderthal DNA, nor do I include published DNA results from many of the Native American skeletons. This differs from the mainstream scientific practice. It also differs from the practice of some creationists. It's fair to ask whether my practice is justified.

My reasons for rejecting the use of ancient DNA are two-fold. First, I've long been suspicious that the DNA is too degraded to be useful. The mainstream community acknowledges the process of degradation. But they also claim to have addressed it, eliminating degraded sequences and publishing only the intact ones.

How do we know the intact ones are valid? How can we be sure that the "intact" sequences are not themselves also the result of some form of degradation? My answer has consistently been that we need an independent test to verify them.

My second reason for rejecting ancient DNA is that I've done an independent test on my own. I've published it.[18] You've also just finished

17 Jeanson 2017.
18 Jeanson 2019.

reading another type of test. In other words, I've spent more than ten years analyzing and testing DNA strictly from living humans and living animal, plant, and fungal species. I've made and tested predictions, and these predictions have worked well. The introduction of ancient DNA into these analyses would dissolve this series of successes.

In other words, if ancient DNA is valid, why can I make so much scientific sense of the world without it? And why does this coherent explanation go away when we introduce ancient DNA? To clarify, my successes are objective—by independent standards. If I can continue to predict the future correctly, why would I change what I'm doing by the introduction of ancient DNA? If ancient DNA is valid, then, yes, my conclusions will need to be altered. But how do we know that ancient DNA is valid? My ability to rigorously explain the world without it suggests that ancient DNA is a poor source of information about the past.

I'll state it another way: My mathematically precise formula to anticipate where, when, and how often in the Y chromosome tree new branches will appear is based on the *exclusion* of ancient DNA. It's a prediction based solely on the DNA from living people. If it keeps working, then why should we trust ancient DNA? If it stops working, then we have cause for revisiting the ancient DNA question. So far, I see no reason to do so.

This concludes my outline of the justification for my genetic conclusions. The following sections describe my framework and logic in other fields relevant to the history of the Americas.

## How to discern among sources for indigenous histories?

My path into exploring indigenous histories began via the *Red Record*. The alignment between the Y chromosome data and the *Red Record* were apparent as early as 2020.[19] At the time, I thought differently about the *Red Record* than I do now. I thought it described an arrival in the Americas in line with the arrival of Y chromosome haplogroup Q. In 2020, I hadn't yet uncovered the precise root of the tree. Back then, my estimates for the arrival of haplogroup Q ranged from the A.D. 200s to 900s. They were wider because they were based on a wide range of tree root positions. This wide range overlapped the

19 Jeanson 2020.

range of arrival dates from the *Red Record* which, at the time, still had several centuries of ambiguity, intrinsic to itself. I hadn't yet resolved it, but the overlap between genetics and indigenous histories was, nonetheless, clear. This result played a significant role in forming my attitude towards other indigenous histories.

I'm aware of a general suspicion toward indigenous histories. I, too, am skeptical of accounts that are often strictly oral, or have unclear provenance, or invoke fantastic or supernatural phenomena. But I can't think of a more egregious example of prematurely rejecting indigenous history than the *Red Record*. It has an entire PhD thesis to its name, dedicated to establishing its fraudulent nature.[20] Genetics shows that the thesis was wrong. Consequently, my thinking ever since has been: If we got the *Red Record* so wrong, how many more indigenous histories have been prematurely rejected?

My current practice for evaluating indigenous histories follows a specific protocol. First, I have limited myself to published literature. I omitted information that I obtained word-of-mouth. This practice ensures I have source material verifiable to anyone. It also prevents any accidental breaking of confidence, should a history have been retold to me solely for private consumption.

Second, I give extra consideration to older accounts. For this book, I've drawn mostly on the literature from the A.D. 1700s, 1800s, and early 1900s. In other words, I focus on the pre-modern era. My fear of more recent publications is that the indigenous histories will have been influenced—consciously or subconsciously—by modern scientific thought. I want to know what Native Americans used to think, not how their histories may have evolved.

Third, I focus on histories that narrate straightforward accounts. I've found many that sound and feel like factual histories. Fantastic, implausible language is minimal. Geographic details are precise. Timeframes are unambiguous. I have more trouble interpreting accounts that fail to meet these criteria.

For example, in 1828, an Iroquoian named David Cusick published *Sketches of the Ancient History of the Six Nations*.[21] It purports

20 Oestreicher 1995.
21 Cusick 1828.

to be an account of the origin and history of the Iroquoian peoples. But Cusick is hesitating in his language, using qualifiers like "perhaps" when describing dates. He also uses seemingly rounded numbers, as if the exact details were lost. I have tried cross-correlating his history with other indigenous accounts, and I can't line his story up. I have not included his narrative in this book.

As another example, numerous Native nations have now-written accounts of "Folk Tales" and other stories. Many involve talking animals or other seemingly fantastic phenomena. I am hopeful that some historical information might still be embedded in these accounts. Perhaps the animals are stand-ins for tribal bands. But, for now, I don't know what to make of these accounts. I have also omitted them from this book.[22]

I will now re-tell the sequence in which I came into indigenous histories. This will also illustrate my methodology to putting the pieces of indigenous histories into a coherent narrative.

As mentioned above, the *Red Record* was the first indigenous history that I was able to verify with genetics. The initial, rough agreement with the haplogroup Q history suggested the document was real. A significant step forward from this point was linking the *Red Record* with haplogroup C. As per the discussion in **chapter 4**, this discovery came about after lining up two data sets: census data for the United States and Canada, and genetic frequency data for Native Y chromosome branches in these countries. This step also resolved the chronological ambiguity internal to the *Red Record*.

This newly resolved absolute chronology (see **Supplement Table A**) for the *Red Record* immediately suggested an obvious correlation with archaeology. Cahokia's archaeological apogee and fall seemed eerily similar to the description given in the *Red Record* (see **chapter 7**). I gained confidence that we were again on the right track.

Shortly thereafter, I found the correlations between the *Red Record* and Algic family linguistics (see **chapter 4** and **Appendix B**). This suggested a natural sequence for the formation of the individual nations within the Algic language family. I then compared this newly derived

22 Except, possibly, where mainstream science has attempted to reconstruct histories based off of them, such as for the nations of the southwest.

sequence to what the individual nations themselves said about their origins. I didn't have a comprehensive set of accounts, but the ones I did have lined up.

At this point, my confidence in the veracity of the *Red Record* grew even more. Genetics, linguistics, and archaeology all seemed to independently agree with the Delaware account.

I broadened my search for indigenous histories in the older ethnological literature. I found more accounts from more nations. By this point, I defaulted to assuming that they were true, so long as they fulfilled the criteria I listed above. My approach seemed to be validated as cross-correlations emerged among the accounts. For example, see the **chapter 5** discussion of the "North Walkers" (*Red Record* reference) and the history of the migrations of the Mescalero Apache.

In several cases, the geography implied by or stated in the indigenous histories also agreed with the geography implied by linguistics. For example, see **chapter 7** for a discussion of the Siouan-Catawban migration histories. See also **chapter 5** for the Athabaskan migrations.

Overall, the deeper I've mined the indigenous histories, the more that cross-disciplinary agreement seems to increase.

This is not to say that all accounts are straightforward to interpret. In the footnotes in previous chapters, I've highlighted some specific areas that have given me pause. For example, some of the ethnologists seem to have been off in their statements about geographic distance. I don't know the explanation for this, but I hesitate to throw out an entire account solely on this basis.

In short, I take each indigenous history on a case-by-case basis. Geographical and mathematical errors give me pause. But cross-disciplinary and cross-tribal agreement increases my confidence that a particular account is real and accurate.

## How does this work challenge existing linguistic models?

In theory, one of the main challenges to the historical reconstruction in the preceding chapters comes from the field of linguistics. Effectively, my model proposes that language diversification happens very rapidly. How else could all the languages in the Algic language family appear in just a few centuries? This contradicts the typical mainstream view that language diversification requires thousands of years.

My model also suggests that entire language families in the Americas formed quickly. For example, I strongly suspect that most of the Contact-era language families were the result of either the haplogroup *Q* or haplogroup *C* migrations. Genetically, the haplogroup *Q* migration looks like a movement of close relatives.[23] Now revisit the **online map**[24] of North America at Contact. I'm suggesting that perhaps more than 50 fundamental language divisions diversified primarily from a single migration event in the A.D. 300s to 600s. Compared to most of the rest of the world, this is an unusual amount.[25]

Prior linguistic work has speculated on ways to re-classify languages in the Americas. Greenberg proposed lumping nearly all language families in North, Central, and South America into a single *Amerind* language family. However, his hypothesis is currently considered invalid.[26]

Nonetheless, I suspect that the classification of American languages is due for a reanalysis. If nothing else, my conclusions suggest a variety of new hypotheses to explore. For example:

(1) Are there links between Old World and New World language families, as per the haplogroup *Q* and *C* migrations? (2) How much influence/borrowing has occurred among and between North American and Mesoamerican languages? The dynamic history of pre-Contact migration in North America and Mesoamerica suggests new candidates for these processes.

I am not a linguist by training. If professional linguists revisited the languages and looked for links along these lines, I wonder what language groupings would look like. I have little difficulty imagining many current language families lumped into larger, new language families.

---

23 I.e., few DNA base pairs separate the deep Native American branches in haplogroup *Q* (see **Supplemental Figure 1** in Jeanson 2022 for details).
If we had access to the same type of data for haplogroup *C*, I suspect that we'd see a similar phenomenon.

24 <answersingenesis.org/go/theyhadnames/>

25 I.e., see the low number of language families in other parts of the world as per Simons and Fennig 2018a, Simons and Fennig 2018c.

26 Campbell 1997, p. 254.

## How does this work challenge existing archaeological models?

My model also challenges (obviously) the mainstream view of North American archaeology. For example, current archaeological thinking does not explain the origin of Mississippian culture via migration from Mesoamerica. It also does not invoke migrations from Asia in the A.D. 300s to 600s or in the A.D. 900s.

I would challenge the mainstream archaeological community to consider whether they've tested all relevant hypotheses. Let's say that migrations from Asia did indeed happen in the A.D. era. What sort of archaeological signature should this event leave? Do we need to see Asian pottery suddenly appear? Asian art and cultural practices? New mounds? We already see a change in mound-building after haplogroup Q arrives. How much indigenous American culture could the new arrivals assimilate? Must they reject all that came before? I think these are important questions that must be revisited in light of the genetics successes I've seen over the last 10 years.

Similar questions apply to the Natchez and Muskogean links to Mesoamerica. Yes, current archaeology seems to reject the hypothesis that the Mississippians and Mesoamericans were in a direct trade relationship. But must migrations result in trade relationships with the home country? Why couldn't the Mississippians have made a clean economic break with Mesoamerica and started anew? I suspect that numerous hypotheses implied by my model have yet to be tested with modern archaeology.

## Summary

My views in this book are driven heavily by the scientific successes I've seen in genetics. I've incorporated indigenous histories, linguistics, and archaeology primarily where I can cross-check them within or across disciplines. I recognize that my conclusions in these latter fields challenge existing views. However, nearly every conclusion I've put in print is, in theory, testable in the future with genetics. For example, we should be able to know very quickly whether the Natchez and Muskogeans migrated from Mexico. Their Y chromosome results will either show a Mesoamerican link or they won't. I'm optimistic that the model I've put forth will be borne out by future genetic discoveries.

# Appendix B: Technical Methods and Documentation

## Documentation of the Seven Population Splits in the *Red Record*

I have listed the following in chronological order of appearance in the *Red Record.*

**First Split**

The text from the *Red Record*:

> After [Constantly On Guard], the sachem was Chilili, the Snow Bird, who spoke of the south.
>
> That our people would be able to grow and spread there.
>
> Southward went Chilili; Eastward went the Beaver.
>
> To Akolaki, Snake Land, southern country, tall pine country, seashore country.
>
> To Eastern country, fishing country, mountain country, game herd country.[1]

In earlier stanzas, "Beaver" was used as a clan name:

> Beaver Head and Great Bird, they said "Let us go!" "To Akomen!" they said.
>
> The allies all declared, "All our enemies will be destroyed."

1 McCutchen 1993, p. 86.

> The northerners agreed, the easterners agreed, across the icy ocean was a better place to be.
>
> On a wondrous sheet of ice all crossed the frozen sea at low tide in the narrows of the ocean.
>
> Ten times a thousand they crossed; all went forth in a night; they crossed to Akomen, the East Land, crossing, marching, marching everyone together.
>
> People of the North, the East, the South; of the Eagle, **of the Beaver**, of the Wolf; the Hunters, Shamans, Headmen; the Women, Daughters, Animals;
>
> All came to settle in the evergreen land. The western people reluctantly came there, for they loved it best in the old Turtle country.[2]

From the context in the first quote above, the Beaver was the splinter group; the main body remained with Chilili.

### Second Split

The text from the *Red Record:*

> The next sachem was Hominy Man; raising crops he began.
>
> The next sachem was the Subdivider; these sachems were helpful.
>
> The next sachem was Shriveled Man; the next sachem was Drought.[3]

At first pass, this section doesn't seem to explicitly state the formation of a splinter group from the main population. However, the above quote represents the English translation of the Delaware text. The *Red Record* also contains pictograms. Let's observe the pictograms for these three stanzas (**Figure 96**).

Notice how the text for the first stanza mentions a single sachem, and the pictogram is solitary. In the third stanza, the text mentions two sachems, and there are two associated pictograms. The second stanza

2 McCutchen 1993, p. 76.
3 Ibid., p. 94.

names a single sachem, but there are still two associated pictograms. This would seem to imply that perhaps there was more than one sachem, perhaps consistent with the surprising use of the plural sachems in "these sachems were helpful." Given the fact that the named sachem was *Subdivider*, I suspect the pictograms represent a split in the population.

### Third Split

The text from the *Red Record:*

> After [Hardened One] was The Denouncer, rebellious and unwilling.
>
> In anger, leaving eastward, some went away in secret.[4]

The main body seems to have remained behind.

### Fourth Split

The text from the *Red Record:*

> East Looking was the sachem, melancholy about the war there was.
>
> 'To the Rising Sun now you must go,' he said; many were those who eastward went.
>
> They separated at the Mississippi; the lazy ones remained behind.[5]

In this case, the main body seems to have crossed the Mississippi.

### Fifth Split

The text from the *Red Record:*

> Little Cloud was the sachem; many were those who left.
>
> The Nanticokes [and] the Shawnee went to the south land.[6]

The main body remained behind.

4 Ibid.
5 McCutchen 1993, p. 102, 106.
6 Ibid., p. 116.

### Sixth Split

The text from the *Red Record:*

> Good Inscribed was the sachem by the great falls.
>
> The eastern people, the Wolf People, moved east and north.[7]

Again, the main body seems to have remained behind.

### Seventh Split

The text from the *Red Record:*

> Quite Ready was the sachem by the shore.
>
> Three were desired, so three were to be the divisions: the Unami Tribe, the Munsee Tribe, the Turkey Tribe.[8]

This is a split in-place—no splinter groups going off in a different direction.

### Ethnic identifications

As per the linguistic discussion in **chapter 4**, we can identify the splinter groups from the *Red Record* with linguistic subdivisions and, therefore, ethnic groups, within the Algic language family. We simply line up the chronology of the language splits with the chronology of the population splits in the *Red Record*. The first *Red Record* split seems to have given rise to the Yurok and Wiyot people. The second, to the Blackfeet. The third spawned the Arapaho-Atsina [i.e., Gros Ventre] and Cree-Montagnais groups. The fourth split moved Cheyenne and Menominee away. In the fifth split, the Core Central Algonquian language group formed, which eventually gave rise to the Fox, Sac, Kickapoo, Mascouten, Shawnee, Miami-Illinois, Ojibwa-Potawatomi-Ottawa, Algonquin (Algonkin), and Saulteaux languages. The sixth and seventh splits birthed the individual members of the Eastern Algonquian language group.

### Additional Notes on Methods

Online Map[9] of North and Central America at Contact

7 Ibid., p. 124.
8 Ibid., p. 136.
9 <answersingenesis.org/go/theyhadnames/>

We redrew the insert map for Contact-era North America in Sturtevant (1996); we also redrew (or drew information from) parts of maps 3, 10, and 12 in Asher and Moseley (2007). These redrawn maps were combined into a single illustration. The maps in Sturtevant (1996) and Asher and Moseley (2007) overlap slightly in the area north of Mexico City. North of this overlap, we went with the reconstruction depicted in Sturtevant (1996). South of this overlap, we went with the reconstruction depicted in Asher and Moseley (2007). On the border, we merged the two maps. Where the displayed area for a particular nation differed for the two maps, we tended to default to redrawing the boundaries with the widest area.

Some names were updated on our redrawn map to maintain consistency between the maps and the text of this book.

In the original map 10 of Asher and Moseley (2007), group 67 is shown both in modern El Salvador and on the Pacific coast of Mexico. In the former, group 67 is shaded green. In the latter, group 67 is shaded pink. We made both areas the same color.

In the original map 12 of Asher and Moseley (2007), the Chibchan and Misumalpan languages are shown as a single language family—Chibcha-Misumalpan. We colored our map so that Chibchan and Misumalpan languages were shown as separate families.

### Maps and tree diagrams for Y chromosome haplogroup distributions

Data, branch diagrams, and maps were taken from Jeanson (2022).

### Calculation of the time of origin for Eskimo-Aleut haplogroup *Q-NWT01* branch

In Jeanson (2022) Supplemental Table 4, I calculated (converted) branch separation times from the Karmin et al. (2015) study. Though Karmin et al. (2015) do not have an Eskimo-Aleut population as part of their study, they do sample the Siberian *Koryaks* living on the Kamchatka Peninsula across the Bering Sea from Alaska (see their Table S1 for the geographic coordinates for the Koryaks).

Four of the Koryak males are found on a branch they labeled haplogroup *Q2b* (as per their Table S6). One *Murut* (of the southeast

Asian country of Brunei) is found on a brother branch known as haplogroup *Q2c*. As per their Table S8, the marker for *Q2* is *Q-F1096*; for *Q2b* is *Q-B143*; for *Q2c* is *Q-F745*. As per Luis et al. (2023) Figure 2, *Q-B143* is downstream of *Q-NWT01*. When I search for *Q-F745* in the FamilyTreeDNA (FTDNA) Discover database (https://discover.familytreedna.com/), *Q-F745* returns *Q-M120*. As per Luis et al. (2023) Figure 2, *Q-M120* is also downstream of *Q-NWT01* and is a brother branch to *Q-B143*.

The Discover database of FTDNA shows that the Murut-Koryak split in Karmin et al. (2015) is a general split between East/Southeast Asian branches and Siberian (and, presumably, Alaskan/Eskimo-Aleut) branches. A search for *Q-NWT01* returns *Q-NGQ11*. These appear to be equivalents. In the FTDNA Discover database, *Q-NGQ11* is downstream of *Q-F1096*. *Q-NGQ11* gives rise to *Q-B143* and *Q-M120*. The *Q-B143* branch leads exclusively to Russians (ethnic groups like the Koryaks are not specified in the FTDNA Discover database); the *Q-M120* branch leads to East, Southeast, South, and Central Asians, but not to Russians. Thus, the split between *Q-B143* and *Q-F745* (*Q-M120*) in the Karmin et al. (2015) tree seems to mark a geographic separation.

The date for this split is the date for node 107 in Karmin et al. (2015) Figure S3. As per Jeanson (2022) Supplemental Table 4, the date for node 107 is somewhere between A.D. 300 and A.D. 636.

Searches at https://discover.familytreedna.com/ were performed on August 8, 2024.

# Appendix C: "I'm a Native American tribal leader; what is your advice on DNA testing?"

I'm glad you asked! Here are several points I made to the Lakota at their 2023 Treaty Council Conference:

## 1. The best path forward for DNA testing is the Y chromosome

Most standard DNA tests examine the DNA that you receive from both parents. The advantage of this approach is that you explore both sides of your family tree. The disadvantage is that each parent dilutes the genetic signal of the other parent.

For example, genetically, 50% of my DNA comes from my mother, 50% from my father.[1] The same principles applied to them when they were conceived. Thus, genetically, each of my four grandparents contributes to only 25% of my DNA. Each of my eight great-grandparents contributes only 12.5%. And so on. In just a few generations, the ancestral signal becomes very hard to detect. Effectively, and for a variety of technical reasons, I don't trust the results that genetic testing companies report when the reported ethnicity level is below 5%.

How can you explore the deeper history of your people? In theory, you'll need to test a type of DNA in which the genetic signal from one parent is not diluted by the genetic signal from the other parent. Two types of DNA fit this criterion: Mitochondrial DNA (inherited

1 These numbers are rough averages. The practical reality of biology means that there are slight deviations from these precise numbers.

through the maternal line) and Y chromosome DNA (inherited through the paternal line). For reasons unknown, the statistics of the mitochondrial DNA-based family tree are poor. It's hard to make precise historical inferences from the data. In contrast, the statistics for the Y chromosome DNA-based tree are much tighter. It's a much better tool to dig deep into the ancestral past.

## 2. DNA testing can be done semi-privately (less expensive) or privately (millions of dollars)

I have found that at least one of the major genetic testing companies is willing to accommodate research projects and the privacy concerns that come with it. For example, if you were to order a test from, say, FamilyTreeDNA (FTDNA) as a regular citizen and not as a research project, you would have to give FTDNA your personal information—name, address, etc. If you were to approach them in hopes of doing a research project on multiple members of your tribe, they might be willing to waive the requirement about personal information. If you can set up an Ethics Review Board to supervise your project, then I think FTDNA might be willing to forego some of their typical requirements. They might supply you with test kits, let you obtain informed consent from your tribal members, let you collect the samples from these tribal members, and then have you return the kits but without disclosing the personal information of the participants. Only you or your designated point person would have that information.

This is the semi-private route.

A more private, but more expensive, route would be to set up your own DNA testing lab. Just a few years ago, this might have involved millions of dollars to set up and expensive maintenance contracts for the equipment. The workflow also would have been inefficient without constant use of the machines. Recent advances in DNA sequencing technology have changed this. You now can do testing much more cheaply and in a modular manner. Oxford Nanopore is one company supplying this new technology. Within a few years' time, there will surely be others doing the same.

The new technology is more expensive than going the route of contracting with a DNA testing company. But it allows you to keep all your samples and results in-house.

## 3. Y chromosome testing has validated indigenous histories

This point is somewhat redundant in light of what this book has shown. But I want to emphasize it as a tool that has already worked. It has a good chance of validating your own tribe's migration history.

## 4. Y chromosome testing can fill in gaps in indigenous history

Under this point, I'm thinking of specific scenarios. For example, consider the indigenous history of the Natchez (see **chapter 6**). Their account contains many details on their migration from Mesoamerica. But it does not seem to identify the details of their pre-migration history. What were they doing in Mesoamerica before leaving for North America? Who were their neighbors? Their relatives? Y chromosome testing might be able to fill in this gap.

Consider also the Salish (**chapter 5**). I have not been able to track down a calendar date for their migrations from the midwest to the west coast. Y chromosome testing might be able to identify the date.

## 5. Inter-tribal collaboration is crucial

Recall a basic biological and genealogical point from **chapter 10**. Mathematically, it's virtually impossible for your tribe to *not* have acquired Y chromosome branches from other tribes. This is simply a consequence of your being human. It's a principle that applies to every people group on earth.

Practically, then, when you do Y chromosome testing of your people, you will almost certainly find a diversity of Y chromosome branches. To figure out what they mean, you will need the aid—the Y chromosome testing results—of other tribes. Some might be close neighbors. Or, given the dynamic history of migrations detailed in this book, others might be quite distant.

## 6. A reasonable place to start is 100 Y chromosome tests

Right now, the amount of information on Native American Y chromosome history is minimal. I'm not aware of any study (yet) that lines up Native American tribes with specific Y chromosome branches. I have attempted for a few tribes in this book. But otherwise, the field is wide open for research. Even 100 men will tell you much history about your own people.

If I can be of help as you navigate this field, you can reach me via: <answersingenesis.org/go/theyhadnames/>

# Appendix D: How did it all begin?

> At the beginning, the sea everywhere covered the earth.
>
> Above extended a swirling cloud, and within it, the Great Spirit moved.
>
> Primordial, everlasting, invisible, omnipresent—the Great Spirit moved.
>
> Bringing forth the sky, the earth, the clouds, the heavens.[1]

This quote isn't from the Book of Genesis. It's not from the Bible. It's the beginning of the *Red Record.*

But surely we can be forgiven for recalling the beginning of another Book:

> *In the beginning God created the heavens and the earth. The earth was without form, and void; and darkness was on the face of the deep. And the Spirit of God was hovering over the face of the waters.*
>
> *Then God said, "Let there be light"; and there was light. And God saw the light, that it was good; and God divided the light from the darkness. God called the light Day, and the darkness He called Night. So the evening and the morning were the first day.*
>
> *Then God said, "Let there be a firmament in the midst of the waters, and let it divide the waters from the waters." Thus God made the firmament, and divided the waters which were under the firmament from the waters which were above*

1 McCutchen 1993, p. 52.

> *the firmament; and it was so. And God called the firmament Heaven. So the evening and the morning were the second day.*[2]

Subsequent to the section that I quoted at the beginning of this chapter, the *Red Record* describes the creation of the first man and woman. It also mentions an evil being—a snake. The *Red Record* tells of a great Flood, though it attributes the cause to the serpent, not to God. In many respects, the *Red Record* sounds a loud echo to the first nine chapters of Genesis.

"But aren't these parallels the result of borrowing from post-Contact European Christian neighbors?" It's the right question to ask. Some indigenous histories do show evidence of this. But I don't think the *Red Record* does. It has too many confirmations in genetics, too many confirmations in other indigenous histories to explain away by Western influence. Therefore, I think it must be authentic and thoroughly indigenous.

If so, then we can ask another question: Why does the *Red Record* parallel Genesis? How did this come about? From a Christian perspective, the answer is straightforward.

After the Flood of Noah's day, God commanded Noah's descendants to spread out. They refused. To force them to spread out, God confused their languages, leading to the formation of the first ethnolinguistic groups.[3]

Before the dispersal, they all spoke one language.[4] They would have shared a common history, a common origins narrative. As they left Babel, each of these new ethnolinguistic groups would have carried the common history and origins narrative with them, in their own language.

Presumably, Noah had knowledge of the Creation, Fall, Flood sequence. Surely his descendants did as well. Yet once they could no longer speak to each other and corroborate their accounts, and if some groups failed to keep written records, the original version could easily have been corrupted in these groups. The *Red Record* attests to

---

2 Gen.1:1-8; NKJV.
3 See Genesis 10:1-11:9.
4 See Genesis 11:1.

its own written documentation late in its own history. In contrast, the Bible implies the existence of written Hebrew records from the earliest times. Hence, the obvious, but imperfect, parallels between the *Red Record* and Genesis.

So…how did the history of North America begin? *When* did it begin? You may have noticed that my timeline for this book starts in 1000 B.C. (e.g., see the heading for **chapter 2**). It's not because I think human history started then. It's because the earliest major culture—Poverty Point—is in full swing by then.[5]

Who came before Poverty Point? When did the earliest Americans set foot in the New World? Did they migrate across the Bering Strait? The answers to these questions come from a variety of disciplines.

To the first question, archaeology has identified a number of cultures. The *Clovis* people are among the earliest. Their spear points (**Figure 97**) have reached almost iconic status.

But when did they live? Mainstream science has long maintained an overall timeline for the history of the Americas going back about 15,000 years. Genetics tells a different story. I've summarized the logic and sequence of the relevant discoveries in **Appendix A**.

To the question of the Bering Strait, I think the answer is yes, the earliest Americans, including the ancestors to the Clovis, could have crossed there. For that matter, they may have crossed the Pacific from Southeast Asia. Either way, I'm of the position that Native peoples arrived in the Americas after a migration from Asia.

Why? As I show in Appendix A, we all go back to Noah and his sons. According to Genesis 6-9, this early family exited the Ark near the mountains of *Ararat*. Whether the modern mountain of the same name is the biblical one, I don't know; the scholarly opinion is divided. But the biblical mountains of Ararat were almost certainly somewhere in the Middle East. Therefore, to arrive in the Americas, surely some body of water—presumably the Pacific, but perhaps the Atlantic—would have been crossed.

Then, at some point, the early arrivals would have built Poverty Point. And the rest, they say, is history.

5 Fagan 2019, p. 196-197.

# Appendix E
# How to contact the author

For video updates on this research, visit my YouTube and Rumble channels:

YouTube: @nathaniel_jeanson

Rumble: NathanielJeanson

You can also find me on social media:

Twitter/X: @NathanielJeans1

Facebook: Nathaniel Jeanson

MeWe: Nathaniel Jeanson

Parler: @Nathanieljeanson

Truth Social: @Nathanieljeanson

Gab: Nathaniel_Jeanson

Telegram: Nathaniel_Jeanson

GETTR: Nathan Jeanson

To contact me directly, you can reach me via the message box at: <answersingenesis.org/go/theyhadnames/>

# Bibliography

*Ames, K.M. and Maschner, H.D.G. 1999. Peoples of the Northwest Coast: Their archaeology and prehistory*. London: Thames and Hudson, Ltd.

Anderson, D.G. and Sassaman, K.E. 2012. *Recent Developments in Southeastern Archaeology: From Colonization to Complexity*. Washington, D.C.: Society for American Archaeology, The SAA Press.

Asher, R.E. and Moseley, C. eds. 2007. *Atlas of the World's Languages*. New York: Routledge.

Beatty, C. 1768. *The Journal of a Two Months Tour*. London: n.p.

Beckwith, C.I. 2009. *Empires of the Silk Road*. Princeton, NJ: Princeton University Press.

Bergström, A. et al. 2020. "Insights into human genetic variation and population history from 929 diverse genomes." *Science* 367(6484):eaay5012.

Bernardini, W., Koyiyumptewa, S.B., Schachner, G., and Kuwanwisiwma, L. 2023. *Becoming Hopi: A History*. Tucson, AZ: The University of Arizona Press.

Bierhorst, J., transl. 1992. *History and Mythology of the Aztecs: The Codex Chimalpopoca*. Tucson, AZ: The University of Arizona Press.

Birch, J. 2015. "Current Research on the Historical Development of Northern Iroquoian Societies." *J. Archaeol. Res.* 23:263-323.

Birmingham, R.A. and Goldstein, L.G. 2005. *Aztalan: Mysteries of an Ancient Indian Town*. Madison, WI: Wisconsin Historical Society Press.

Birmingham, R.A. and Rosebrough, A.L. 2017. *Indian Mounds of Wisconsin*. Madison, WI: The University of Wisconsin Press.

Brinton, D.G. 1885. *The Lenape and Their Legends; with the Complete Text and Symbols of the Walam Olum, a New Translation, and an Inquiry into Its Authenticity*. Philadelphia, PA: D.G. Brinton.

Campbell, L. 1997. *American Indian Languages: The Historical Linguistics of Native America*. New York: Oxford University Press.

Clark, W.P. 1885. *The Indian Sign Language, with Brief Explanatory Notes of the Gestures Taught Deaf-Mutes in our Institutions for their Instruction*. Philadelphia: L.R. Hamersly & Co.

Clark, A.J. and Bamforth, D.B., eds. 2018. *Archaeological Perspectives on Warfare on the Great Plains*. Louisville, CO: University Press of Colorado.

Coe, M.D. and Houston, S. 2015. *The Maya*. New York: Thames & Hudson Inc.

Coe, M.D. and Koontz, R. 2013. *Mexico: From the Olmecs to the Aztecs*. New York: Thames & Hudson Inc.

Crow, Joseph Medicine. 1992. *From the Heart of the Crow Country: The Crow Indians' Own Stories*. New York: Orion Books.

Cusick, D. 1828. *Sketches of the Ancient History of the Six Nations*. Tuscarora Village: Lewiston, Niagara Co.

de Milford, Louis LeClerc. (G. de Courcy, transl.; J.F. McDermott, ed.). 1956. (Orig.: Published 1802) *Memoir or A Cursory Glance at My Different Travels & My Sojourn in the Creek Nation*. Chicago: The Lakeside Press.

Dorsey, J.O. 1886. "Migrations of Siouan Tribes." *The American Naturalist* 20(3):211-222.

Du Pratz, M. Le Page. 1774. *The History of Louisiana or of the Western Parts of Virginia and Carolina*. (transl. from the French). London: T. Becket.

Dulik, M.C. et al. 2012. "Y-chromosome analysis reveals genetic divergence and new founding native lineages in Athapaskan- and Eskimoan-speaking populations." *PNAS* 109(22):8471-8476.

Eldredge, N. 1982. *The Monkey Business: A Scientist Looks at Creationism*. New York: Washington Square Press.

Fagan, B. 2019. *Ancient North America: The archaeology of a continent*. New York: Thames & Hudson.

Futuyma, D. and Kirkpatrick, M. 2017. *Evolution*. Sunderland, MA: Sinauer Associates, Inc.

Gatschet, Albert S. 1884. *A Migration Legend of the Creek Indians, with a Linguistic, Historic and Ethnographic Introduction, Vol. 1*. Philadelphia: D. G. Brinton.

Goodman, R. (Seeger, A., editor.) 2017. *Lakota Star Knowledge: Studies in Lakota stellar theology*. Mission, SD: Sinte Gleska University Press.

Greenberg, J.H. 1987. *Language in the Americas*. Stanford, CA: Stanford University Press.

Gregory, D.A. and Wilcox, D.R., eds. 2007. *Zuni Origins: Towards a new synthesis of Southwestern archaeology*. Tucson, AZ: The University of Arizona Press.

Grinnell, George B. 1904. *Pawnee Hero Stories and Folk Tales with Notes on the Origin, Customs and Character of the Pawnee People*. New York: Charles Scribner's Sons.

Grinnell, George B. 1923. *The Cheyenne Indians: Their History and Way of Life, Vol. 1*. New Haven: Yale University Press.

Haywood, John. 1823. *The Natural and Aboriginal History of Tennesee up to the First Settlements Therein by the White People in the Year 1768*. Nashville: George Wilson.

Healan, D.M., Cobean, R.H., and Bowsher, R.T. 2021. "Revised chronology and settlement history of Tula and the Tula region." *Ancient Mesoamerica* 32:165-186.

Heckewelder, J. 1819. "An account of the history, manners, customs, of the Indian Natives who once inhabited Pennsylvania and the Neighboring States," in: *Transactions of the Historical & Literary Committee of the American Philosophical Society*. Philadelphia: Abraham Small.

Hill, J.B. et al. 2004. "Prehistoric Demography in the Southwest: Migration, Coalescence, and Hohokam Population Decline." *American Antiquity* 69(4):689-716.

Hill Jr., M.E. and Ritterbush, L.W., eds. 2022. *People in a Sea of Grass: Archaeology's Changing Perspective on Indigenous Plains Communities*. Salt Lake City, UT: The University of Utah Press.

Holcombe, C. 2017. *A History of East Asia*. Cambridge, UK: Cambridge University Press.

Hudson, C. and Tesser, C.C. 1994. *The Forgotten Centuries: Indians and Europeans in the American South*. Athens, GA: The University of Georgia Press.

Inomata, T. et al. 2020. "Monumental architecture at Aguada Fénix and the rise of Maya civilization." *Nature* 582(7813):530-533.

Jeanson, N.T. 2013. "Recent, functionally diverse origin for mitochondrial genes from ~2700 metazoan species." *Answers Research Journal* 6:467–501.

Jeanson, N.T. 2015a. "Mitochondrial DNA clocks imply linear speciation rates within 'kinds.'" *Answers Research Journal* 8:273-304.

Jeanson, N.T. 2015b. "A young-earth creation human mitochondrial DNA 'clock': Whole mitochondrial genome mutation rate confirms D-loop results." *Answers Research Journal* 8:375-378.

Jeanson, N.T. 2016. "On the Origin of Human Mitochondrial DNA Differences, New Generation Time Data Both Suggest a Unified Young-Earth Creation Model and Challenge the Evolutionary Out-of-Africa Model." *Answers Research Journal* 9:123-130.

Jeanson, N.T. 2017. *Replacing Darwin: The New Origin of Species*. Green Forest, AR: Master Books.

Jeanson, N.T. 2019. "Testing the Predictions of the Young-Earth Y Chromosome Molecular Clock: Population Growth Curves Confirm the Recent Origin of Human Y Chromosome Differences." *Answers Research Journal* 12:405-423.

Jeanson, N.T. 2020. "Young-Earth Y Chromosome Clocks Confirm Known Post-Columbian Amerindian Population History and Suggest

Pre-Columbian Population Replacement in the Americas." *Answers Research Journal* 13:23-33.

Jeanson, N.T. 2022. *Traced: Human DNA's Big Surprise*. Green Forest, AR: Master Books.

Jeanson, N.T. and Holland, A.D. 2019. "Evidence for a Human Y Chromosome Molecular Clock: Pedigree-Based Mutation Rates Suggest a 4,500-Year History for Human Paternal Inheritance." *Answers Research Journal* 12:393-404.

Jeanson, N.T. and Lisle, J. 2016. "On the Origin of Eukaryotic Species' Genotypic and Phenotypic Diversity: Genetic Clocks, Population Growth Curves, and Comparative Nuclear Genome Analyses Suggest Created Heterozygosity in Combination with Natural Processes as a Major Mechanism." *Answers Research Journal* 9:81-122.

Karmin, M. et al. 2015. "A recent bottleneck of Y chromosome diversity coincides with a global change in culture." *Genome Res.* 25(4):459-466.

Lapham, I.A. 1855. *The Antiquities of Wisconsin: As surveyed and described*. New York: G.P. Putnam & co. [also listed as published in Washington City by the Smithsonian Institution]

Lekson, S.H. 2015. *The Chaco Meridian: One thousand years of political and religious power in the ancient Southwest*. Lanham, MD: Rowman & Littlefield.

Luis, J.R., Palencia-Madrid, L, Garcia-Bertrand, R., and Herrera, R.J. 2023. "Bidirectional dispersals during the peopling of the North American Arctic." *Scientific Reports* 13:1268.

Lynott, Mark J. 2014. *Hopewell Ceremonial Landscapes of Ohio: More than mounds and geometric earthworks*. Havertown, PA: Oxbow Books.

Mann, C. 2005. *1491*. New York: Alfred A. Knopf.

Mason, O.K. and Friesen, T.M. 2017. *Out of the Cold: Archaeology on the Arctic Rim of North America*. Washington, DC: The SAA Press.

McCutchen, D., trans. 1993. *The Red Record: The Wallam Olum*. Garden City Park, NY: Avery Publishing.

McEvedy, C. and Jones, R. 1978. *Atlas of World Population History*. Middlesex, England: Penguin Books.

*McLean v. Arkansas Board of Education*, https://law.justia.com/cases/federal/district-courts/FSupp/529/1255/2354824/.

McNutt, C.H. and Parish, R.M., eds. 2020. *Cahokia in Context: Hegemony and Diaspora*. Gainesville, FL: University of Florida Press.

Mills, B.J. and Fowles, S., eds. 2017. *The Oxford Handbook of Southwest Archaeology*. New York: Oxford University Press.

Milner, G.P. 2021. *The Moundbuilders: Ancient Societies of Eastern North America*. New York: Thames & Hudson, Inc.

Mooney, James. 1896. "The Ghost-Dance Religion and the Sioux Outbreak of 1890." *Fourteenth Annual Report of the Bureau of Ethnology, part 2*. Washington: Government Printing Office.

Mooney, James. 1898. "Calendar History of the Kiowa Indians", extract from *The Seventeenth Annual Report of the Bureau of American Ethnology*. Washington: Government Printing Office.

Moorehead, W.K. and Laufer, B. 1922. *The Hopewell mound group of Ohio*. Field Museum of Natural History, Publication 211, Chicago. Vol.VI, No.5.

N.A. 1954. *Walam Olum or Red Score: The Migration Legend of the Lenni Lenape or Delaware Indians; a new translation, interpreted by linguistic, historical, archaeological, ethnological, and physical anthropological studies*. Indianapolis, IN: Indiana Historical Society.

Neitzel, J.E., ed. 1999. *Great Towns and Regional Polities: In the Prehistoric American Southwest and Southeast*. Albuquerque, NM: University of New Mexico Press.

Nichols, D.L. and Pool, C.A., eds. 2012. *The Oxford Handbook of Mesoamerican Archaeology*. New York: Oxford University Press.

Oestreicher, D.M. 1995. *The Anatomy of the Walam Olum: A 19th Century Anthropological Hoax*. Ph.D. dissertation, Rutgers University. New Brunswick, New Jersey.

Olofsson, J.K., Pereira, V., Børsting, C., and Morling, N. 2015. "Peopling of the North Circumpolar Region—Insights from Y Chromosome STR and SNP Typing of Greenlanders." *PLoS One* 10(1): e0116573. doi:10.1371/journal.pone.0116573.

Oneroad, Amos and Skinner, Alanson B. (Anderson, L.L., ed.). 2003. *Being Dakota: Tales and Traditions of the Sisseton and Wahpeton.* St. Paul, MN: Minnesota Historical Society Press.

Ortman, S.G. 2012. *Winds from the North: Tewa origins and historical anthropology.* Salt Lake City, UT: The University of Utah Press.

Pauketat, T.R., ed. 2012. *The Oxford Handbook of North American Archaeology.* New York: Oxford University Press.

Pauketat, T.R., and Sassman, K.E. 2020. *The Archaeology of Ancient North America.* Cambridge, UK: Cambridge University Press.

Philbrick, N. 2020. *Mayflower: A Story of Voyage, Community, and War.* (no city): Penguin Books.

Plog, S. 2008. *Ancient Peoples of the American Southwest.* London: Thames & Hudson.

Pinotti, T. et al. 2019. "Y Chromosome Sequences Reveal a Short Beringian Standstill, Rapid Expansion, and early Population structure of Native American Founders." *Curr. Biol.* 29(1):149-157.

Poznik, G.D. et al. 2016. "Punctuated bursts in human male demography inferred from 1,244 worldwide Y-chromosome sequences." *Nat. Genet.* 48(6):593-599.

Raff, J. 2022. *Origin: A Genetic History of the Americas.* New York: Twelve, Hatchette Book Group.

Russell, F. 1975. *The Pima Indians.* Tucson, AZ: The University of Arizona Press.

Sarmiento de Gamboa, Pedro. 1572. *History of the Incas.* Translated by Sir Clements Markham K.C.B. (1907). https://www.yorku.ca/inpar/sarmiento_markham.pdf

Schoolcraft, H.R. 1851. *Historical and Statistical Information Respecting the History, Condition and Prospects of the Indian Tribes of the United States: Part 1.* Philadelphia: Lippincott, Grambo & Company.

Schoolcraft, H.R. 1855. *Information Respecting the History, Condition and Prospects of the Indian Tribes of the United States: Part V*. Philadelphia: J.B. Lippincott & Company.

Seymour, Deni J., ed. 2012. *From the Land of Ever Winter to the American Southwest: Athabaskan Migrations, Mobility, and Ethnogenesis*. Salt Lake City: The University of Utah Press.

Shackley, M.S. 2019. "The Patayan and Hohokam: A view from Alta and Baja California." *Journal of Arizona Archaeology* 6(2):83-98.

Simons, G.F. and Fennig, C.D., eds. 2018a. *Ethnologue: Languages of Asia*. Dallas, TX: SIL International Publications.

Simons, G.F. and Fennig, C.D., eds. 2018b. *Ethnologue: Languages of the Americas and the Pacific*. Dallas, TX: SIL International Publications.

Simons, G.F. and Fennig, C.D., eds. 2018c. *Ethnologue: Languages of Africa and Europe*. Dallas, TX: SIL International Publications.

Squier, E.G. and Davis, E.H. 1848. *Ancient Monuments of the Mississippi Valley*. New York: Bartlett & Welford.

Steponaitis, V.P. and Scarry, C.M., eds. 2016. *Rethinking Moundville and Its Hinterland*. Gainesville, FL: University Press of Florida.

Sturtevant, William C. 1979. *Handbook of North American Indians, Vol.9, Southwest*. Washington: Smithsonian Institution.

Sturtevant, William C. 1996. *Handbook of North American Indians, Vol.17, Languages*. Washington: Smithsonian Institution.

Sturtevant, William C. 2001. *Handbook of North American Indians, Vol.13, Part 1 of 2, Plains*. Washington: Smithsonian Institution.

Swann, B., ed. 2005. *Algonquian Spirit: Contemporary Translations of the Algonquian Literatures of North America*. Lincoln, NE: University of Nebraska Press.

Teit, James A., Farrand, L., Gould, M.K., and Spinden, H.J. (edited by Franz Boas). 1917. *Folk-Tales of Salishan and Sahaptin Tribes*. Lancaster, PA, and New York: The American Folk-lore Society.

Thwaites, R.G. (ed.). 1906. *Early Western Travels 1748-1846*. Cleveland, Ohio: The Arthur H. Clark Company.

Trabert, S.J. and Hollenback, K.L. 2021. *Archaeological Narratives of the North American Great Plains: From Ancient Pasts to Historic Resettlement*. Washington, D.C.: Society for American Archaeology, The SAA Press.

Warren, W.W. 1885. *History of the Ojibway Nation*. Saint Paul, MN: Minnesota Historical Society.

Webster, D. 2002. *The Fall of the Ancient Maya: Solving the Mystery of the Maya Collapse*. New York: Thames & Hudson Inc.

Williams, Samuel Cole, ed. 2022. *Adair's History of the American Indians*. https://www.gutenberg.org/files/67699/67699-h/67699-h.htm (Original: Adair, James. 1775. *The History of the American Indians*. London: Edward and Charles Dilly.)

Wingard, J.D. and Hayes, S.E., eds. 2013. *Soils, Climate & Society: Archaeological Investigations in Ancient America*. Boulder, CO: University Press of Colorado.

# Credits and Sources for Photos and Illustrations

**Map of Native Languages of North and Central America also available as a large poster for sale through www.masterbooks.com.**

Online Map of North and Central America at Contact:
Redrawn from Sturtevant 1996 map insert at the back of the volume, and from Asher and Moseley 2007 Maps 3, 10, and 12.

**Figure 1** — Redrawn from Squier and Davis map, 1848 (for original see: https://commons.wikimedia.org/wiki/File:Newark_Works_Squier_and_Davis_Plate_XXV.jpg)

**Figure 2** — Getty Images, #1500525896, credit: Mary Salen.

**Figure 3** — Getty Images, #1473458055, credit: Mary Salen.

**Figure 4** — Source: Jubileejourney, https://commons.wikimedia.org/wiki/File:Octagon_Earthwork.JPG

**Figure 5** — Redrawn from images at http://ancientohiotrail.org/sites/newark and http://www.earthworks.uc.edu/

**Figure 6** — Getty Images, #1416298240, credit: Mary Salen.

**Figure 7** — Redrawn from Ephraim George Squier and Edwin Hamilton Davis, https://commons.wikimedia.org/wiki/File:Hopeton_Work_Squier_and_Davis_Plate_XVII.jpg

**Figure 8** — Redrawn from Berthold Laufer and Warren King Moorehead, https://commons.wikimedia.org/wiki/File:The_Hopewell_mound_group_of_Ohio_BHL6714907.jpg

**Figure 9** — Adapted from Figure 9.4 (p.317) of Pauketat and Sassman (2020).

**Figure 10** — Hopewell core area adapted from Figure 9.4 (p.317) of Pauketat and Sassman (2020).

**Figure 11** — Redrawn and adapted from Figure 9.13 (p.333) of Pauketat and Sassman (2020). With inspiration from Herb Roe, https://commons.wikimedia.org/wiki/File:Poverty_Point_Aerial_HRoe_2014.jpg

**Figure 12** — Getty Images, #900704140, credit: Rodrigo Pardo.

**Figure 13** — Adapted from plate 18, p.40, of Coe and Koontz (2013).

**Figure 14** — Source: Alfonsobouchot, https://commons.wikimedia.org/wiki/File:La_Venta_Pir%C3%A1mide_cara_sur.jpg

**Figure 15** — Adapted from Figure 34.3, p. 413, of Boszhardt, R. "The Effigy Mound to Oneota Revolution in the Upper Mississippi River Valley," in: Pauketat (2012).

**Figure 16** — Redrawn from Increase Lapham, https://commons.wikimedia.org/wiki/File:Dewey_Mound_Group_Map_Lapham.jpg

**Figure 17** — Source: Jeanson 2022, Color Plate 22.

**Figure 18** — Source: Jeanson 2022, Color Plate 187.

**Figure 19** — Modified from Jeanson 2022, Color Plate 194.

**Figure 20** — Adapted from plate 1, p. 12, of Coe and Houston (2015).

**Figure 21** — Getty Images, #600418188, credit: SL_Photography.

**Figure 22** — Getty Images, #501453380, credit: f9photos.

**Figure 23** — Source: Jeanson 2022, Color Plate 26.

**Figure 24** — Extracted from Online Map of North and Central America at Contact (see above).

**Figure 25** — Extracted and adapted from Online Map of North and Central America at Contact (see above).

**Figure 26** — Redrawn from McCutchen (1993).

**Figure 27** — Source: Jeanson 2022, Color Plate 188

**Figure 28** — Extracted and adapted from Online Map of North and Central America at Contact (see above).

**Figure 29** — Extracted and adapted from Online Map of North and Central America at Contact (see above).

**Figure 30** — Extracted and adapted from Online Map of North and Central America at Contact (see above).

**Figure 31** — Extracted and adapted from Online Map of North and Central America at Contact (see above).

**Figure 32** — Extracted and adapted from Online Map of North and Central America at Contact (see above).

**Figure 33** — Redrawn from: https://gaftp.epa.gov/EPADataCommons/ORD/Ecoregions/cec_na/NA_LEVEL_I.pdf https://www.epa.gov/eco-research/ecoregions-north-america

**Figure 34** — Extracted and adapted from Online Map of North and Central America at Contact (see above).

**Figure 35** — Extracted and adapted from Online Map of North and Central America at Contact (see above).

**Figure 36** — Extracted and adapted from Online Map of North and Central America at Contact (see above).

**Figure 37** — Extracted and adapted from Online Map of North and Central America at Contact (see above).

**Figure 38** — Extracted and adapted from Online Map of North and Central America at contact (see above).

**Figure 39** — Extracted and adapted from Online Map of North and Central America at Contact (see above).

**Figure 40** — Getty Images, #1183298374, credit: MattGush.

**Figure 41** — Getty Images, #990712260, credit: MattGush.

**Figure 42** — Getty Images, #175431120, credit: toddmedia.

**Figure 43** — Getty Images, #1087398112, credit: rodclementphotography.

**Figure 44** — Getty Images, #481272289, credit: JoseIgnacioSoto.

**Figure 45** — Credit: Diego Rivera, https://commons.wikimedia.org/wiki/File:Templo_Mayor_in_Mexico-Tenochtitlan_16th_century_(illustration_1900).jpg

**Figure 46** — Extracted and adapted from Online Map of North and Central America at contact (see above).

**Figure 47** — Extracted and adapted from Online Map of North and Central America at contact (see above).

**Figure 48** — Extracted and adapted from Online Map of North and Central America at contact (see above).

**Figure 49** — Redrawn and adapted from: https://www.epa.gov/sites/default/files/2015-03/marb_600x395.jpg

**Figure 50** — Extracted and adapted from Online Map of North and Central America at Contact (see above).

**Figure 51** — Extracted and adapted from Online Map of North and Central America at contact (see above).

**Figure 52** — Redrawn from Figure 15.5, p. 226, of Anderson, D.G., "Examining Chiefdoms in the Southeast: An Application of Multiscalar Analysis," in: Neitzel (1999).

**Figure 53** — Extracted and adapted from Online Map of North and Central America at contact (see above).

**Figure 54** — Redrawn and adapted from Figure 1, p. 76-77, of Hudson, C., "The Hernando de Soto Expedition, 1539-1543," in: Hudson and Tesser (1994).

**Figure 55** — Getty Images, #1187722894, credit: zrfphoto.

**Figure 56** — Redrawn and adapted from: https://cdn.britannica.com/44/244344-050-9271A018/Locator-map-Illinois-River.jpg

**Figure 57** — Redrawn and adapted from Figure 3.1, p. 62, of Meeks, S.C. and Anderson, D.G., "Drought, Subsistence Stress, and Population Dynamics: Assessing Mississippian Abandonment of the Vacant Quarter," in: Wingard and Hayes (2013).

**Figure 58** — Extracted and adapted from Online Map of North and Central America at Contact (see above).

**Figure 59** — Extracted and adapted from Online Map of North and Central America at Contact (see above).

**Figure 60** — Extracted and adapted from Online Map of North and Central America at Contact (see above).

**Figure 61** — Redrawn and adapted from Figure 34.4, p. 415, of Boszhardt, R. "The Effigy Mound to Oneota Revolution in the Upper Mississippi River Valley," in: Pauketat (2012). Also from Figure 4-2, p. 93, of Trabert and Hollenback (2021).

**Figure 62** — Extracted and adapted from Online Map of North and Central America at Contact (see above).

**Figure 63** — Extracted and adapted from Online Map of North and Central America at Contact (see above).

**Figure 64** — Extracted and adapted from Online Map of North and Central America at Contact (see above).

**Figure 65** — Extracted and adapted from Online Map of North and Central America at Contact (see above).

**Figure 66** — Extracted and adapted from Online Map of North and Central America at contact (see above).

**Figure 67** — Extracted and adapted from Online Map of North and Central America at Contact (see above).

**Figure 68** — Extracted and adapted from Online Map of North and Central America at Contact (see above).

**Figure 69** — Getty Images, #1163157440, credit: Danielle Leonard.

**Figure 70** — Extracted and adapted from Online Map of North and Central America at Contact (see above).

**Figure 71** — Getty Images, #1312999141, credit: Greg Meland.

**Figure 72** — Getty Images, #1206279876, credit: David Arment.

**Figure 73** — Getty Images, #2099912986, credit: JeffGoulden.

**Figure 74** — Getty Images, #2163994978, credit: Billy McDonald.

**Figure 75** — Getty Images, #159232044, credit: alexeys.

**Figure 76** — Getty Images, #636676866, credit: Jeff Goulden.

**Figure 77** — Getty Images, #2152226158, credit: Pgiam.

**Figure 78** — Getty Images, #1433949933, credit: milehightraveler.

**Figure 79** — Getty Images, #609946562, credit: kojihirano.

**Figure 80** — Redrawn and adapted from Figure 10.5, p. 230, of Fagan (2019).

**Figure 81** — Getty Images, #175226951, credit: powerofforever.

**Figure 82** — Getty Images, #1188044452, credit: scgerding.

**Figure 83** — Redrawn and adapted from Figure 49.1, p. 600, of Lekson, S.H., "Chaco's Hinterlands," in: Pauketat (2012).

**Figure 84** — Redrawn and adapted from Figure 9.1, p. 201, of Fagan (2019).

**Figure 85** — Redrawn from Plate 93, p.112, of Plog (2008).

**Figure 86** — Extracted and adapted from Figure 84 (see above).

**Figure 87** — Redrawn and adapted from Figure 1.4 of Hopkins et al., "Hopi Perspectives on History," p. 19, in: Bernardini et al. (2023).

**Figure 88** — Getty Images, #508956914, credit: Legate

**Figure 89** — Extracted and adapted from Online Map of North and Central America at Contact (see above).

**Figure 90** — Adapted and redrawn from Figure 19.2 of Ferguson, T.J., "Zuni Traditional History and Cultural Geography," p. 384, in: Gregory and Wilcox 2007.

**Figure 91** — Redrawn from Figure 3 of Hill et al. 2004.

**Figure 92** — Extracted and adapted from Online Map of North and Central America at Contact (see above).

**Figure 93** — Getty Images, #1282534573, credit: miralex.

**Figure 94** — Source: Jeanson 2022, Color Plate 12.

**Figure 95** — Source: Jeanson 2022, Color Plate 13.

**Figure 96** — Redrawn from McCutchen (1993).

**Figure 97** — Getty Images, #105558449, credit: Brian_Brockman.